P9-DUC-545

BEYOND THE BOTTOM LINE
Measuring World Class Performance

**The Cooper and Lybrand
Performance Solutions Series**

**Linking Theory and Practice
to Develop Unique Solutions
for Contemporary Performance Issues**

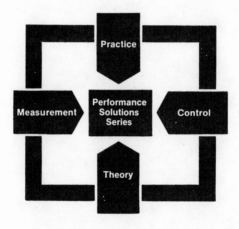

Other Titles in This Series
*Crossroads: A JIT Success
Story*
Carol J. McNair and Robert Stasey

**Carol J. McNair
Series Editor**

BEYOND THE BOTTOM LINE
Measuring World Class Performance

Carol J. McNair
William Mosconi
Thomas Norris

Dow Jones-Irwin
Homewood, Illinois 60430

Sponsoring editor: Jim Childs
Project editor: Joan A. Hopkins
Production manager: Carma W. Fazio
Compositor: Weimer Typesetting Company, Inc.
Typeface: 11/13 Century Schoolbook
Printer: R. R. Donnelley & Sons Company

Library of Congress Cataloging-in-Publication Data

McNair, Carol Jean.
    Beyond the bottom line: measuring world class performance/Carol
J. McNair, William Mosconi and Thomas Norris.
        p.   cm.
    Includes index.
    ISBN 1-55623-194-6
    1. Manufactures—Accounting. 2. Inventories—Accounting. 3. Just
-in-time system—Accounting. 4. Just-in-time systems—Evaluation.
5. Efficiency, Industrial—Evaluation. I. Mosconi, William.
II. Title.
HF5686.M3M38 1989
657'.867—dc19   89—1063
                                                                   CIP

Printed in the United States of America
1 2 3 4 5 6 7 8 9 0 D 6 5 4 3 2 1 0 9

To our family and friends who supported us throughout this project

ACKNOWLEDGMENTS

In an undertaking of this magnitude, there are always a great number of people who make it work. First and foremost, Patrick Romano and Alfred King of the National Association of Accountants deserve a heartfelt thanks for their work on the original project and for the willingness to let the material from this study be published in its current form. This appreciation, of course, extends beyond these two individuals to the association itself and the funding so generously provided under the auspices of the Bold Step Project.

In addition to this group, numerous individuals have read the drafts and provided helpful comments along the way. Gordon Shillinglaw, professor of accounting at Columbia University, has continuously provided support and constructive criticism of the manuscript. Professor William Rotch's comments were the basis for improving several key arguments in the book. Professor Anthony Hopwood pushed to move beyond the obvious. Others who read and critiqued the drafts include Professor H. Thomas Johnson, Robert Howell, Thomas Soucey, and an anonymous reviewer.

Finally, a special thank you goes to each of the individuals at the companies visited who gave generously of their time and expertise, providing an honest overview, and detail of their company's efforts to achieve excellence in the competitive global marketplace. Without these individuals, as well as those cited above, this book could not have been completed.

CONTENTS

INTRODUCTION

Just-in-time manufacturing (JIT) is one of several new process design approaches surfacing in practice. Ranging from simple inventory control techniques to complex systems that integrate the entire production process from inception to delivery, these technologies are creating a demand for new forms of accounting and control information. The practicing accountant is being bombarded with a series of requests for new information, for increased relevance of reports and measures, and for recognition that various technologies or manufacturing practices may require changes in the basic accounting system.

JIT: A FOCUS ON TIME

The basic theme of JIT is to have *only* the right part in the right place at the right time. It is a "pull" system of manufacturing, with material movements being triggered solely by customer demand. In this sense, the next workstation is the customer. No demand—no movement or production.

What sets JIT-based manufacturing apart from more traditional repetitive or job shop processes is its emphasis on the flow of materials through the plant and its focus on time as the basic variable in efforts to improve the production process. Specifically, *cost is caused* by events that use time. This relationship, at the heart of the current move to identify 'cost drivers,' represents a shift from using the management accounting system solely to track incurred costs.

Group technology is the key to implementing a JIT approach to manufacturing because it emphasizes product families rather than functional process groupings. For example, whereas a traditional manufacturing facility is organized by function (e.g., rows of punch presses followed by rows of lathes), floor designs based on group technology place machines together according to the path followed by a specific set of products. In other words, group technology clusters dissimilar machines based on the common assembly path of a product family. This change in plant layout suggests a new type of cost pool that resembles the machine-labor-hour concepts first envisioned by A. Hamilton Church in 1929.

Reducing Work-in-Process

Inventory is the primary enemy of JIT manufacturing. Logically, reorganizing the plant floor into cells reduces work-in-process inventory, as there is little room for staging components and materials between machines. This explains why current literature emphasizes eliminating these inventory buffers.

Another system supporting this inventory reduction is *Kanban,* a Japanese term meaning card. Kanban creates a 'pull' manufacturing system by limiting the amount of production at prior stages to that contained in the Kanban square, or tray. Kanban technologies provide a way to visually control the manufacturing process without costly computerized tracking systems. Focusing on minimizing WIP inventories, Kanban attacks the *move* and *queue* elements of the production process. Two implications for management accounting systems emerge from this system: increased visibility of the production process and a means to identify the value-added components of the process.

Increasing Emphasis on Prevention

Preventing errors is another message embedded in the JIT philosophy. By introducing preventive maintenance and total quality control, JIT minimizes unplanned disruptions of the production process. Preventive maintenance reduces line stoppage due to machine breakdowns during periods of planned production.

Quality, or the lack thereof, is another source of disruption in the system which JIT manufacturing tries to control. In the interdependent system described above, quality problems can shut down an entire plant. While at first glance this seems undesirable, the alternative produces mass quantities of waste, rework, or scrap that remains undetected and uncorrected. From a costing standpoint, quality control, set-up, and maintenance—formerly considered to be indirect, or support, activities—are transformed into value-added functions directly tied to the manufacturing process.

Summing Up JIT

This brief introduction to the basic technologies and practices generally associated with JIT manufacturing serves as a backdrop to the rest of the book. Simplicity is an overriding characteristic of this system. A simplicity gained through discipline and common sense forms the core of JIT's philosophy, or way of doing business, one that focuses on the continual elimination of waste. It is based on process-flow redesign in order to minimize the time elapsed between the receipt of a customer order and the delivery of the item to the customer's door. Waste in this setting is defined as the nonproductive use of time—whether in reference to idle inventory or time spent reworking defective components. Figure A summarizes these points, providing a convenient way to integrate the materials into your own organization or work.

FIGURE A
Minimizing Interval

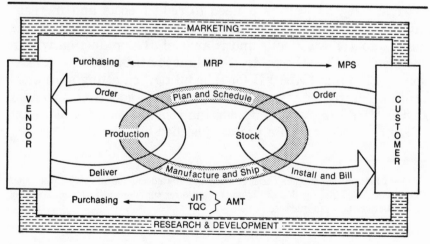

THE COMPANIES IN THIS BOOK

Five primary sites, visited from October 1986 to February 1987, provided a rich and varied basis for the discussions in this book. Actual field visits ranged from one to three days per site, during which time the three principal researchers conducted in-depth, semistructured interviews with key personnel. The research instrument used is reproduced at the end of this book (see Appendixes A and B).

Diesel Systems, Inc.

Diesel Systems, Inc.,[1] is a $337 million enterprise that designs, manufactures, and markets complex precision metal products and components for foreign and domestic industrial and consumer markets. Its main products are diesel fuel injectors, nozzles, fuel heaters, mechanical tappets, and a range of related products used to produce diesel engines.

Diesel Systems, Inc., produces 800 automotive (e.g., 25 models) and 100 agricultural fuel injectors (over 3,000 models) per day at the site studied. Revenues generated from these sales exceeded $215 million in 1987. This production level is maintained using approximately 30 percent of the existing plant capacity.

Due to downturns in the domestic diesel market, Diesel Systems, Inc., is facing a period of declining profits and sales. Overcapacity, as well as a need to reduce costs and increase quality, has provided the impetus for the conversion of the plant into JIT work cells and islands of JIT production within the plant studied. Currently, the plant is utilizing Material Resource Planning (MRP) II for scheduling; installing statistical process control where possible; gradually adopting a vendor certification program; and adopting robotics and flexible manufacturing systems on a process-justified basis.

---

[1]These fictitious names for the organizations visited are maintained throughout this discussion. This was done to honor requests for confidentiality from several of the participating companies.

TeleComm Corporation

TeleComm Corporation is a multibillion-dollar electronics and telecommunication enterprise. Although the majority of its revenue comes from the domestic market, it also operates manufacturing and distribution facilities throughout the United States, Europe, and the Far East.

TeleComm is composed of several core businesses: long-distance services, communications equipment for office and home use, and network telecommunications equipment. The plant studied during this research project produces personal computer components, such as printers, keyboards, display terminals (CRTs), and controllers. The plant occupies 695,000 square feet and employs 1,300 direct laborers and 350 expense, or indirect, individuals. Production at this facility was about $288 million in 1987.

A slowdown in the growth of the computer industry as it moves into maturity has created the need to control costs and standardize products. A JIT line was developed and partially installed within this plant, creating a true factory-within-a-factory setting complete with performance measurement systems and incentives. MRP II is utilized plantwide, vendor certification on a corporate level is moving forward, and robotics are employed at various locations throughout the plant.

CompSci Industries

CompSci Industries is a multibillion-dollar organization that operates in a single industry segment: the design, manufacture, and service of measurement and computation products and services. It is expanding its sales and production network into the international market and is broadening its involvement in joint ventures in Europe, the Far East, and South America.

Two CompSci plants were visited during the course of this study, both of which produce personal computer componentry, such as tape drives, memory, and printers. Most of our analysis focused on the tape drive assembly plant. This site has the most highly developed JIT lines, as well as advanced MRP II systems, and has made significant changes to its Management Accounting System (MAS) to match the characteristics of JIT manufacturing. Cycle time has been reduced from an average of 22 days to 1 day, overhead per unit has been decreased 30 percent, and a two-tiered overhead system has been devised to reduce the distortions present in utilizing a direct labor base in a JIT setting.

MicroChip, Inc.

MicroChip, Inc., is one of the world's leading manufacturers of electronic equipment, systems, and components. Its extensive, complex product line spans four major product groups: communications, semiconductors, information systems, and government electronics. The site studied is part of the semiconductor products sector of MicroChip, Inc. It produces integrated circuits, custom logic boards, and related semiconductor technology-based products. Its sales were approximately $1.9 billion in 1987, or roughly 30 percent of the corporate total. This is up 9 percent from 1986, with new orders rising 42 percent and backlogs decreasing 3 percent.

This site currently is changing its manufacturing processes to a "focused-factory" approach, utilizing work cells to replace traditional functional departments. To date, 29 cost blocks have been converted to 11 work cells. The primary focus in the measurement area is to reduce cycle time by improving the linearity of production, and to reduce inventories, scrap, and quality problems. Cell costs are computed as an hourly rate, and are measured against a theoretical optimum rather than a standard.

General Business, Inc.

General Business, Inc., is a large multinational producer of office equipment, computers, and related items. It boasts multi-billion-dollar annual sales levels worldwide, roughly half of which comes from international markets. Its corporate structure reflects this worldwide marketing approach, with a separate subsidiary focusing on sales to various world regions. General Business's primary objectives for 1987 and beyond include strengthening and renewing the product line, significantly increasing the number of people directly serving customers, and sharpening the focus on the customer and his or her needs.

The General Business site visited during this study is a fully automated, computer-integrated assembly plant producing laptop computers. Utilizing 13 robots and no direct labor, each of the two Computer-Integrated-Manufacturing (CIM) lines in this plant can turn out one computer every two minutes.

At this site, accounting measurement has become routinized and automated, described best as a "no-brainer" situation. Its primary emphasis is on zero defect production, decreased cycle times, and development of a direct charging accounting system that will seek to charge to a product the actual overhead incurred. The basic criterion in the MAS is the degree to which the process/person adds value to the final product.

Getting Started

This book examines the new technologies emerging in manufacturing and the far-reaching changes their use is having on accounting systems. As the details unfold, the fragmented, and unique, path each company is taking to excellence may surprise you. Reality is that these technologies require far-reaching organizational changes that do not appear overnight. Each company has been able to make some progress—none is anywhere near the end of the journey.

Where this new wave of change will end cannot be guessed. But change will occur across all industries. By describing how leading-edge companies are getting started, the gates to the future are opened. Success lies somewhere down the path to excellence—beyond the barriers of accounting and organizational traditions.

CHAPTER 1

A TURBULENT ENVIRONMENT

Today's manufacturing environment in the United States is turbulent and laced with uncertainty. Competition from abroad is constantly eroding both market share and profits of major American corporations. These organizations are being pushed to the wall: They must change their manufacturing practices and regain a competitive edge in their markets or face entering the ranks of the displaced.

The challenges confronting manufacturers both here and abroad are changing the rules of the game. To be successful, companies can no longer compete on a single dimension, such as cost. Instead, they must excel at two or more of the traditional strategic elements. Low cost, high quality, and high customer service levels are the basic ingredients of success in this unsettling environment.

Managers seeking to prosper in this environment are turning to their management accounting system for new types of information, such as product life cycle accounting. These new information demands require a proactive mentality, both on the plant floor and in the controller's office.

A Process of Dynamic Change

A brief glimpse at newspapers and business magazines leaves even the uninitiated with a feeling of overwhelming frustration. Toffler's *Future Shock* warns of the increasing pace of change, but few of us are prepared to live with it. In the business sector, managers are faced not with one new technology, but with an evolutionary and perhaps revolutionary process of dynamic change.

These changes can be traced to increased competition in the world marketplace, shortened life cycles for most products, and the impact of the computer on all aspects of modern business. By eliminating waste and adopting flexible approaches to product and process design, productivity improvements can provide the means for maintaining or regaining a competitive edge in this turbulent environment.

"Simplify, automate, integrate" are the bywords of the movement to advanced manufacturing technologies (AMT). They embody the concept of continuous improvement, or constant change, that typifies today's manufacturing environment. As suggested by Figure 1–1, manufacturing excellence is the goal being pursued. Advanced technologies represent one path to achieving this goal.

FIGURE 1–1
The Pursuit of Excellence

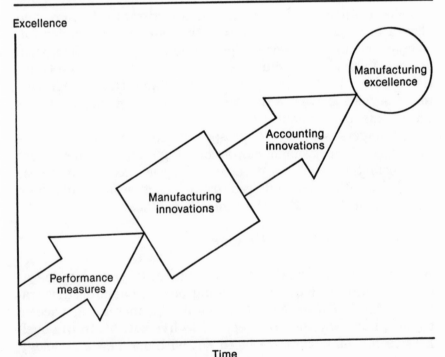

Advanced manufacturing technologies exploit the computer's information-processing and problem-solving capabilities to enhance and improve manufacturing processes and control mechanisms. For managers trained in the complexities of Gantt charts and the merits of net present value calculations, the implications of computer-aided design, numerical control machines, and the peopleless factory are difficult to grasp. Yet beneath the confusion, there is a new excitement among operating managers. As technology enters to refocus their activities from "putting out fires" to plotting the strategic role of manufacturing, managers are both accepting and embracing these challenges. Instead of desperation, one senses determination, and from determination comes success.

JIT Manufacturing: A Journey to Excellence

Just-in-time [JIT] manufacturing, a comprehensive approach based on the concept of continuous improvement, is one of the new tools being used by managers to meet the challenges of their competitive environment. JIT technologies focus on eliminating waste and improving product flow from inception to delivery. JIT and its benefits provide the backdrop for this research.

JIT manufacturing is dedicated to producing the right parts at the right time—when they are *needed* rather than when they can be made. It is a pull manufacturing system, which moves goods through a shop based on end-unit demand. Its goal is to maintain a constant flow of components and product, rather than batches of work-in-process (WIP) inventory. This goal contrasts sharply with the traditional push system, which moves goods when they are finished rather than when they are needed. The objective is to reduce, and finally eliminate, work-in-process and the problems and costs it hides.

JIT is not just a manufacturing process, it is a journey. It is a companywide philosophy that focuses on reducing waste and controlling the production process. JIT counsels management to change its attitude toward inventory by recognizing that a large WIP inventory does not represent job security, but instead leads to unemployment, quality problems, higher operating costs, and lower profits.

An overriding theme in a JIT environment is continuous improvement and change. The objective is to simplify by removing the complexities inherent in the manufacturing processes of the 60s and 70s. JIT is, after all, a commonsense approach to manufacturing that requires a change in the organization's mindset.

The benefits of JIT are related to reductions of time, space, and effort (e.g., inventory). Reducing inventory removes the buffers from the manufacturing process and exposes production problems previously hidden. In a JIT environment, a company becomes acutely aware of the fact that quality improvements reduce rather than increase costs. Defects stop the line, result in missed deliveries, create new costs through rework, necessitate employment of vast numbers of quality assurance personnel, and harm product sales if not detected. Focusing on doing the right job the first time decreases costs, safely reduces inventories, and enhances product attributes.

JIT AND THE MANAGEMENT ACCOUNTING SYSTEM

The fact that management accounting has basically not changed in form and procedures since the late 1920s is reason enough to question its usefulness today. Yet, the gap between manufacturing practices and the accounting system used to control it developed gradually. Why, then, is the field suddenly at a crisis point? The answer is quite simple: Advanced technologies amplify existing problems. Just as they isolate the waste and inefficiencies in the physical production process, they also highlight the discrepancies inherent in our management accounting procedures. James Brimson, in a recent article, notes:

> The problem facing most manufacturers is that their facilities are not structured to meet these demands, and there are many roadblocks that make the transition to an automated factory difficult. One of the most important but least understood of these roadblocks is current cost management systems. These systems

do not provide companies with the financial information necessary to manage the transition to a factory of the future.[1]

One major shortcoming of traditional management accounting systems [MAS] springs from an overemphasis on control to the exclusion of support for planning. Managers looking over their shoulders at the failures of yesterday fail to anticipate the demands of tomorrow. And cost accountants, trained to serve those managers, as well as safeguard the financial integrity of the organization, have not taken the initiative to change the information or services they provide. This book argues that accountants must adopt a proactive mentality, one that advocates the process of continuous improvement. Only by discarding the current emphasis on control in favor of the proactive approach can accountants truly support their company's strategic manufacturing goals.

Another major area of concern noted in the popular press is the integrity of the information produced by our management accounting systems (e.g., product costs). Many authors suggest that automation of manufacturing processes has drastically changed both the types of costs incurred by firms and the relationships among these costs. Traditional management accounting techniques build upon the assumption that direct labor represents the primary component of product costs. Overhead rates are computed and allocated based on these direct labor costs.

Yet in today's environment, direct labor represents an increasingly insignificant part of product cost. In a fully automated, computerized setting (e.g., computer-integrated manufacturing) traditional direct labor practically disappears from product costs, as assemblers are replaced by machines, machine tenders by robots and computers, and floor supervisors by computer programmers. Because these changes in the components of product cost cannot be captured by existing cost accounting models, they result in distortions, frustration, and distrust between accountants and management.

---

[1]J. Brimson, "How Advanced Manufacturing Technologies Are Reshaping Cost Management," *Management Accounting*, March 1986, p. 25.

THE MANAGEMENT ACCOUNTING SYSTEM—REALITY

The MAS, as currently structured, is designed for an environment containing few products and few work centers. The traditional full-cost absorption accounting methods, which apply standards by products and variances by department or work center, provide an adequate information base in this setting.

At the time of its development, this financial accounting-oriented system was probably the most cost-efficient tool for tracking product costs and meeting the demands of external reporting. The problems with this system can be traced to a more complex manufacturing environment, which now supports many products passing through multiple work centers and departments. Increased complexity requires an improved and expanded information flow. Global variances by functions and work centers cannot be relied on to track the complex causal factors that drive costs.

Interestingly, the pursuit of simplicity in process is highlighting the distortions created by a mismatched MAS. When the production environment was unnecessarily complex, it was difficult to isolate the specific weaknesses in the information system [e.g., MAS]. The long-standing friction between production and accounting indicate that these distortions in the accounting system are not new. In fact, the MAS's loss of relevance was a gradual process. Repairing it, though, cannot be as gradual, because the rapidly changing environment of the future will require relevance, speed, and relative accuracy of performance information. To meet these demands, the MAS must simplify its accountings to enhance its capacity to process information.

In this book we use multiple case studies to answer the basic research question: What types of changes are occurring in the management accounting system in response to advanced manufacturing (e.g., JIT) technologies?

Five primary sites provided a rich and varied basis for this study: Diesel Systems, Inc.; TeleComm Corporation; CompSci Industries; MicroChip, Inc.; and General Business, Inc. Actual field visits ranged from one to three days per site, during which the three principal researchers conducted in-depth, semi-structured interviews with key personnel. Site visits were conducted from October 1986 to February 1987. The research instrument is reproduced in the appendix to this book.

Brief Description of Research Sites

Diesel Systems, Inc., designs, manufactures, and markets complex precision metal products and components for foreign and domestic industrial and consumer markets. Its main products are diesel fuel injectors, nozzles, fuel heaters, mechanical tappets, and a range of related products used in the production of diesel engines. Due to downturns in the domestic diesel market, Diesel Systems is facing a period of declining profits and sales. Overcapacity, as well as a need to reduce costs and increase quality, provided the impetus for converting the plant into JIT work cells and islands of JIT production. Currently, the company is using MRP II for scheduling, installing statistical process control where possible, gradually adopting a vendor certification program, and adopting robotics and flexible manufacturing systems on a process-justified basis.

TeleComm Corporation is an electronics and telecommunication enterprise. The particular plant used in this study produces printers, keyboards, display terminals, and controllers. A slowdown in the growth of the computer industry as it matures has created the need to control costs and standardize products in an attempt to reduce production costs. A JIT line was developed and partially installed within this plant, creating a true factory-within-a-factory setting complete with performance measurement systems and incentives. MRP II is utilized plantwide, vendor certification on a corporate level is moving forward, and robotics are employed at various plant locations.

CompSci Industries is a leader in the electronics industry. Two plants were visited during the course of this study, both of which produce personal computer componentry, such as tape drives, memory, and printers. The majority of the analysis focused on the tape drive assembly plant. This site had the most highly developed JIT lines, as well as advanced MRP II systems, and has made significant changes to its MAS to match the characteristics of JIT manufacturing. Cycle time has been reduced from an average of 22 days to 1 day, overhead per unit has been decreased 30 percent, and a two-tiered overhead system has been devised to reduce the distortions present in utilizing a direct labor base in a JIT setting.

MicroChip, Inc., manufactures integrated circuits in the site studied. It currently is changing its manufacturing processes to a focused factory approach, utilizing work cells to replace traditional functional departments. To date, 29 cost blocks have been converted to 11 work cells. Their primary focus is to reduce cycle time by improving the linearity of production and to reduce inventories, scrap, and quality problems. Cell costs are computed on an hourly basis, which is then compared against a theoretical optimum rather than a standard.

General Business, Inc., is a large multinational producer of office equipment, computers, and related items. The site used in this study builds typewriters for the domestic market utilizing a computer-integrated manufacturing approach. At this site, accounting measurement has become routinized and automated, described best as a "no-brainer" situation. Its primary emphasis is on zero defect production, decreased cycle times, and development of a "direct charging" accounting system that will charge to a product the actual overhead incurred. The basic criterion in the MAS is the degree to which the process/person adds value to the final product.

TRENDS IN MAS

Several major observations about the current manufacturing environment (detailed in subsequent chapters) can be made:

1. The current JIT environment is typified by islands of innovation, both in manufacturing technology and performance measurement (MAS) change.
2. The absolute amount of change is still very small, both within individual firms and across the manufacturing sector.

3. The management accounting profession is following, not leading, the process of change and is often inhibiting rather than instigating productive manufacturing change.
4. To properly serve management, the MAS needs to be the source of different costs for different purposes.
5. "Being approximately right" is a keynote in a value-added approach to MAS redesign.
6. One main problem is the failure of accountants to "stick to their knitting." Specifically, traceability must once again become the primary criterion in developing information to support product costing, control, coordination, planning, and related MAS cost objectives.

The future of manufacturing competitiveness lies in the rapid development of JIT and other AMT principles on and off the shop floor. As labor and material costs even out throughout the world's markets, the value-added use of overhead will provide a competitive edge. The lessons embedded in the concepts of interval reduction and elimination of non value-added activities must be applied in research and development, engineering, accounting, order processing, distribution, and other traditional staff functions. Attendant with these efforts will be the reduction of management hierarchies, followed by an increase in the velocity of decision making. As suggested by A. Hamilton Church, the glob called *overhead* needs to become our friend in the pursuit of efficient, effective manufacturing.

MAS REDESIGN

JIT accounting entails many major changes in the policies, procedures, and systems used to design and operate the MAS. Three primary trends differentiate a JIT accounting model from traditional MAS approaches: (1) JIT costing approximates process accounting, (2) actuals replace standards as the primary basis for applying costs, and (3) volume-based costs are replaced by an increased focus on traceability.

Data First

In an environment where flexibility and quality replace cost as the primary criterion for success, the basic assumptions and relationships monitored by the MAS must be changed. Relevant events encompass far more than pure economic transactions in a traditional accounting model.

The basic methods and analyses developed to address management accounting issues are not irrelevant or extinct. Rather, it is the data on which these techniques are based that need to be restructured. The data-collection tools and techniques employed by the management accountant affect the relevance and accuracy of subsequent reports. If the data lack integrity, the entire process becomes a garbage in–garbage out exercise. Fully allocated, historical accounting-based product costs are not the sole purpose for maintaining the MAS.

Major Change Ahead

Current management accounting systems require more than a few adjustments. In fact, cost-based reporting systems should be expanded to measure total performance whether operational or financial, quantitative or qualitative. It will necessitate an expanded role for the management accountant, one that includes providing a full menu of data for management decision making. This expanded role will remove the financial accounting orientation that has dominated cost accounting, allowing the management accountant to assume a proactive, vital role in the management of the organization.

CHAPTER 2

TECHNOLOGY: PROVIDING A SOLUTION

Improvements in technology and procedures are associated with higher standards of living in the United States. The ability to innovate is perceived as one of our major strengths. We rely on Yankee ingenuity to solve problems and provide us with a competitive edge in the world marketplace.

Ironically, our current lack of competitiveness is due in large part to technological improvements in manufacturing processes and procedures. Richard Schonberger suggests this turnabout can be traced to a postwar complacency that turned the task of running a manufacturing enterprise into gentlemen's work. Manufacturing firms are no longer run by the experienced manager who dedicates his career to a single entity. Instead, people two and three times removed from the manufacturing process man the helm.

But today this scene is changing again. Managers are beginning once more to manage their companies rather than the numbers. Many are doing so to survive. Recognizing that the new manufacturing technologies can provide a competitive edge, these managers are dramatically changing their production processes. This chapter details the key changes occurring in the manufacturing environment, suggesting that manufacturing excellence arises from the proactive adoption of advanced manufacturing technologies and philosophies.

UNDERSTANDING THE NEW ENVIRONMENT

The manufacturing environment of the 1990s will differ significantly from that of the 80s. Figure 2–1 details six major areas of change: process and facilities, computer hardware, planning and control, product design, financial control, and organization. No aspect of the environment will be untouched by the onslaught of technological advancements.

Processes and Facilities

A typical manufacturing plant is best characterized by one term: *redundancy.* Multiple discrete machines requiring mul-

FIGURE 2–1
The New Manufacturing Environment

Features and Functions	Current Manufacturing Environment	Manufacturing Environment of the 90s
Process and Facilities	• Many discrete machines • Multiple set-ups • Large warehouses • Large WIP areas	• Flexible machine centers • Zero set-up • No warehouses • Drastic decline in space required
Computer Hardware	• Main frame • Mini • Micro/PC	• Mini ⟶ Mainframe Micro ⟶ ↓ PC ⟶ Multiple networks
Planning and Control	• Constant demand fluctuation • Infinite rescheduling of requirements • Constant engineering change • Weekly planning • Long lead times • Large lot sizes • Vendor difficulties	• Demand stabilization • Minimum rescheduling • Zero change • Hourly planning • Zero lead times • Lot size of 1 • Vendor synergies
Product Design	• Life cycle declining • Constant engineering change • Many complex components • Quality improvement over cycle • Infinite options	• Life cycle much shorter • Little or no engineering change • Few complex components • 100% quality at job #1 • Limited options
Financial Control	• Labor efficiency • Little emphasis on investment • Shop orientation • Focus on variable cost • Overhead spreading • Cost measurement	• Product profitability full stream • Investment intensive • Product cost as incurred • Minimum variable cost beyond material • Zero direct labor • Cost, flexibility, dependability and quality measures
Organization	• Functional interfaces • Long lead times • Hierarchical	• Product teams • Flexible and rapid decision making • Fewer levels

tiple, time-consuming set-ups buffered by large inventories stored in large warehouses decouple the manufacturing process from demand, while ways are sought to capitalize on economies of scale. A focus on efficiency dominates effectiveness, quality, and competitive concerns.

Contrast this description to the projection for the 1990s. Flexible machine centers allow the firm to reduce capital intensity, increase responsiveness to customer demands, and provide the advantages of economies of scope (e.g., variety). Flexible manufacturing, when coupled with the concept of zero set-up time, eliminates inventory buffers. The result: no warehouses. This combination of reduced inventory and fewer, more flexible machines requires much less space. The total picture is one of leanness: one machine is used for several products/processes, wasting no resources in non–value-adding set-ups and yielding competitive advantages through both enhanced responsiveness and reduced costs.

Planning and Control

The current manufacturing environment reflects the struggle between a dynamic, changing marketplace and the quest for stability in the manufacturing process. Artificially constructed production schedules lead to infinite rescheduling and the appearance of constant demand fluctuation. They create long lead times and encourage production in large lot sizes to achieve economies of scale. Add to this vendor difficulties, and the result is lack of control.

In sharp contrast, the flexibility and responsiveness obtained via advanced manufacturing technologies [AMT] stabilizes demand. As lot sizes of one become feasible, rescheduling practically disappears, lead times go to zero, and planning becomes possible on an hourly basis. Planning and control in the manufacturing environment of the 1990s will be a "no-brainer" as the production process becomes the only buffer needed between a dynamic environment and the organization. In such a setting, vendor synergy is both a result of and a prerequisite to effective operations.

Financial Controls

Current accounting systems are dedicated to bean counting: most cost measurement techniques focus on labor efficiencies, overhead spreading, and identification of variable costs. Because massive resources are dedicated to tracking costs incurred in the shop, investment support for process or product improvements is neglected. In this setting, the management accountant is little more than an extension of his or her calculator, a scorecard rather than an active part of the management team.

An AMT environment allows management accountants to turn in their green eyeshades and calculators. Product profitability becomes the main focus, characterized by full-stream tracking of costs as incurred from inception to retirement. Another major area of activity, investment decision support and analysis, places accountants on the project development team.

As direct labor continues to decrease as a component of product cost, the impact of overhead allocations based on direct labor increases. In many respects, direct labor is becoming a fixed cost. Improved traceability will help reinstitute the boundaries between direct and indirect costs, resulting in fewer *arbitrary* allocations. Thus, the narrow focus on cost measurement is being replaced by a total performance measurement approach that monitors flexibility, dependability, quality, and cost.

Complementary changes are occurring in computer hardware and software applications, product design, and organizational structure and functioning. In each area, buffers (or waste) are removed as the responsiveness of the system increases.

Four Key Trends

Four key trends, or areas of change, result from the adoption of advanced manufacturing technologies. First, the *basis of competition changes* in an AMT setting, requiring the use of multidimensional manufacturing strategies. Second, *the pace of change accelerates,* as reflected in shorter product life cycles and more frequent engineering change. Third, *the flow of materials through the plant replaces bottleneck capacity* as the key criterion of manufacturing efficiency. Finally, the *pattern of costs in the factory changes* from variable to fixed, and direct to indirect.

NEW TECHNOLOGIES AND POTENTIALS FOR SUCCESS

The increasing number of new terms used to describe and identify the various AMTs can create confusion among the uninitiated. This section will briefly define key terms and discuss the characteristics of the various technologies.

MRP and MRP II

MRP, or materials requirement planning, was one of the first technologies to hit the plant floor in the United States. The objectives of MRP are: (1) planned and controlled inventories, in particular having items available at the time of usage, and (2) generation and revision of replenishment actions on an item-by-item basis. MRP is a materials scheduling procedure geared toward maintaining adequate inventory levels.

MRP is based on a master production schedule (MPS), which details the timing (e.g., order release dates) and quantities of production of all final products. A bill of materials (BOM) provides a precise list of the components required, by product, and their quantities. This procedure integrates scheduling and material handling.

When MRP is expanded to include capacity planning, shop floor control, and purchasing, it is called MRP II, or manufacturing resource planning. Originally envisioned as an integrated manufacturing system, it was designed to provide manufacturing managers with a unified database for control and planning, as well as a "what if" capability. Currently, though, the term *MRP II* is used interchangeably with *MRP,* both indicating a scheduling module dedicated to planning production and material acquisitions. MRP is a basic form of automation that provides stability in a traditional manufacturing environment.

The adoption of MRP, and the scheduling and planning discipline it entails, is now widespread in U.S. manufacturing firms. While resulting in inventory reductions and enhanced basic control processes in manufacturing organizations, it is not the MRP II dreamed of by Ollie White, its creator. Rather than serving as a tool to integrate the entire plant, it is a back office function detailing the location and movement of inventory.

CAE, CAD, CAM

MRP systems provided the first step in factory computerization. Not far behind came utilizing the computer to design products. Computer-aided design (CAD) and computer-aided engineering (CAE) are growing in use, nudging the world of product design into the age of technology. Computer-aided manufacturing [CAM] is closing the loop of computer-based control/planning tools.

Computer-aided design uses the computer to standardize and improve productivity in the designing, drafting, and testing of products. Because a CAD system can analyze a tremendous number of dimensions and alternative configurations, it facilitates the comparison of various product and process designs on the basis of cost, simplicity, and materials.

CAD and CAE provide the ability to design products utilizing existing components, using the CAD database to identify and incorporate standard parts in new products. Both of these technologies speed up the task of designing products and restrict the growth of part numbers. Using proven parts translates to decreased costs and improved responsiveness. New products become available faster and with fewer failures. As componentry is standardized, inventory management is simplified and the threat of obsolescence is reduced. CAD and CAE provide the tools to achieve "design for manufacturability."

They simplify product design and reduce the total number of active parts in inventory.

Computer-aided manufacturing technologies "use the computer to plan, implement, and control the production of a product through optimal utilization of existing manufacturing facilities and resources."[1] The term covers several forms of computer-based manufacturing processes, ranging from a single machine to an entire plant.

UTILIZING COMMON SENSE IN MANUFACTURING

KANBAN Systems

Kanban systems are physical locations (bins, racks, or pallets) placed between workstations. They provide a visible signal that an in-process unit can be passed to the next worker. The system developed by Hewlett-Packard is often cited as a premier Kanban site.

Within a JIT cell, production is paced by the use of Kanban outsquares, physical controls that prevent inventory buildup at any stage in the production process. This constraint reduces work-in-process inventory and enhances quality. If a part does not fit or a test is not passed, the entire line stops. Kanban systems provide a visible, simple tool for process control that can be implemented without the computer. The only requirement is common sense.

Kanban systems convert repetitive manufacturing to process manufacturing. Product flow is visible, constant, and free of buffers and waste, so work orders and work-in-process inventory accounts are eliminated. Because a Kanban system utilizes a pull rather than push approach to moving products through the manufacturing process, parts buildup is limited to

---

[1]R. Schonberger, *World Class Manufacturing: The Lessons of Simplicity Applied,* (New York: The Free Press, 1986).

what can be contained within the Kanban square. The Kanban concept represents a valuable tool for process control in a JIT environment.

Finally, Kanban concepts dramatically shorten set-up times. Once again, a change in mindset and the use of common sense distinguish the improvements made under a Kanban system from computer-based approaches. The objective is to make set-up time insignificant. Minimizing changeover costs increases the profitability of small lot sizes.

Total Quality Control

Total quality control (TQC) is based on a simple concept: Do it right the first time. It is usually associated with statistical process control (SPC), a data-intensive procedure that collects information on the critical factors in a manufacturing process. SPC signals the need to shut down a process if any measured factor falls outside of the acceptable range. SPC is not a new concept. Before the revolution on the factory floor, it was just another tool of the quality control specialist. Today, however, this basic control technique is used directly on the plant floor and is yielding dramatic improvements in quality.

SPC uses Gantt charts to track the key characteristics of the production process. At regular intervals, a small sample of output is obtained, predetermined critical factors are measured, and the average of the readings is recorded. If this number falls between the upper and lower tolerance levels for that factor, no action is taken. If, however, the item falls outside of the acceptable range, the process is stopped and the problem corrected. As in a Kanban system, problems in the production process are identified and isolated when they occur rather than later when additional processing costs have been incurred.

By using SPC and related techniques, production's focus shifts from inspecting defects out of the finished goods to building in quality at each stage of the process. Contrary to prior beliefs, improving quality actually reduces, rather than increases, product costs. Costly rework is minimized, waste and scrap eliminated, and inventories reduced.

ATTACKING SET-UPS

Each of the techniques described above focuses on a single aspect of the production process, resulting in improved control and reduced inventory costs. Another way to shorten cycle time is to attack set-ups. This brings the discussion to numerical control devices: a series of programmable tooling and production devices, otherwise called *unitary machines*.

Numerical Control

Numerical control (NC) systems include a broad range of programmable machine tools. This class of machines can be programmed with a set of instructions that serve as an autopilot, guiding the performance of desired machining operations.

The term NC often denotes early numerical control machines that use a punched paper tape to store the various set-up and machining instructions. Because specific jobs are coded in a retrievable, reusable form, flexibility is gained. When jobs are changed, a new set of instructions (e.g., tape) is loaded. If a new job is run, simple changes are made to an existing program. Thus, NC concepts, much like Kanban, focus on reducing non–value-added time.

Using the Computer—CNC and DNC

When a computer is utilized in the programming phase, the process becomes computer numerical control (CNC). CNC uses a dedicated computer in or beside the NC machine. In place of the paper tape, computer programs store the various configurations and perform the machining operations. Specific programs can be created or modified directly on the machine, eliminating the problems associated with reliance on paper tapes as a mode of storage and retrieval.

Direct numerical control (DNC) goes one step beyond CNC, linking a group of NC machines to one central computer. There is the enhanced potential for data collection in a DNC shop; the systems gather feedback from each machine on production rates and machine status, as well as provide instructions for specific machining operations.

The entire group of NC technologies provides many benefits, such as: (1) productivity gains of up to 300 percent; (2) reduced set-up time; (3) improved quality and reduced variability in output; (4) decreased scrap and rework levels; and (5) marked reductions in the number of direct labor workers needed to achieve a desired level of output.

FLEXIBLE MANUFACTURING: A NEW MANUFACTURING ENVIRONMENT

Each of the above technologies enhances the efficiency of existing processes and can be implemented in a traditional manufacturing environment. A traditional manufacturing environment is organized along functional lines and is dominated by a direct labor orientation. Machines and information systems are utilized to improve worker productivity rather than replace him or her.

In contrast, three alternative manufacturing approaches are surfacing in a new environment best denoted as *flexible manufacturing:* Just-in-time, islands of automation, and computer-integrated manufacturing. Creating flexibility and responsiveness by reducing set-up *times,* and achieving first-pass quality, these three manufacturing approaches typify the "simplify, automate, integrate" view of modern manufacturing.

These new approaches build upon the concept of group technology—a clustering of machines based on production, rather than marketing, families. As defined by Schonberger, "A production family is a group of parts that follow the same flow path."[2]

Cellular manufacturing, an offshoot of the general group technology concept, describes a miniproduction line that minimizes queue and move-time by the sequence and proximate arrangement of machines or people. Often conceptualized as a U-shaped loop, it can assume any basic form. Elimination of space between processing steps and sequential movement of

---

[2]Ibid., p. 10.

product through the area characterize a cellular manufacturing approach.

Two related techniques also focus on creating clusters of production on the plant floor: JIT production and islands of automation (IA). JIT production is the embodiment of simplification and elimination of waste on the plant floor. JIT denotes cellular manufacturing supported by vendor management and logistics improvements that seek to minimize queue and move time and, in the process, inventories.

IA is the automation of various functions or cells, reflecting the second stage of the simplify-automate-integrate cycle.

Finally, computer-integrated manufacturing (CIM) is an integrated, plantwide automated system of manufacturing controlled by a central processing unit. IA and CIM are discussed briefly below to close the loop on existing technologies. The discussion then focuses on JIT manufacturing.

Islands of Automation

Islands of automation consist of an integrated collection of automated production processes, a materials transport system, and one or more controlling computers, which are combined into a system dedicated to the manufacture of a variety of products. While both JIT and IA focus on the production of a family of products with common characteristics, IA dictates the intensive use of computerized processes and controls. IA is often called *flexible manufacturing,* but this term includes the entire family of technologies that minimize set-up time in order to drive inventories to zero.

IAs employ various advanced manufacturing technologies in concert to optimize the performance of subgroups of machines or processes dedicated to specific product families. Robotics are one tool common to IAs. Pick-and-place robots move items from an automated transfer cart to a desired machine or process. A central computer controls and coordinates the automated procedures.

Where JIT is a model of simplicity, IA requires almost total automation of the production process. JIT implementation is a low-cost solution to manufacturing problems and inefficiencies; IAs require a large capital investment. Because technology costs replace direct labor costs in IA settings, IAs represent a major step on the road to a peopleless factory.

Computer-Integrated Manufacturing

The final step toward automation is represented by the concept of CIM. As described by Brimson, "CIM ties together the seemingly diverse threads of the manufacturing enterprise by providing an automatic link among product design, manufacturing engineering, and the factory floor".[3] CIM links the various islands of automation into one integrated system that optimizes the performance of an entire factory.

CIM production is complex. A computerized control system links the entire manufacturing process, and robots perform the majority of assembly and production procedures. The system is software constrained; it is only as flexible as the programs that run it. While production control is complex, product design is simplified in a CIM environment. Because robots can perform only a limited number of procedures, product designers focus on minimizing manufacturing complexity.

The shift in costs from labor to technology is complete in a CIM environment, which automates many indirect labor functions, such as order entry. Given the predominance of fixed, indirect cost, volume has a major impact on final product costs in a CIM environment. CIM is one alternative for a firm whose goal is to achieve manufacturing excellence.

JIT: A JOURNEY, NOT A TECHNIQUE

JIT is a companywide philosophy of doing business, rather than a manufacturing process for improving efficiency on the plant floor. Based on the goal of continuous improvement in all production and nonproduction functions, JIT focuses on producing the right quantity of component parts and subassemblies just as they are needed for use on the parent assemblies. Much like Kanban, it is based on a pull rather than a push

---

[3]J. Brimson, "CIM: Vision or Illusion", *Business Software Review*, April 1987 p. 42.

approach to manufacturing. Goods are pulled through a shop, maintaining a constant flow of materials. JIT procedures allow the tight coupling of manufacturing processes.

JIT techniques keep materials moving—when needed and over the shortest distance. By eliminating the space between workstations and moving machines into manufacturing cells by product groups, non–value-added time is minimized. These changes streamline material flow and reduce the need, as well as the space available, for WIP inventory.

JIT manufacturing techniques effectively reduce the feasible lot size to one, shorten lead times, provide an environment for attaining zero defects, and result in tremendous cost reductions. All of this is achieved without loss of variety. It provides for management by sight, exposing problems rather than hiding them. It represents ultimate simplicity in production.

A Philosophy of Management

JIT manufacturing is best described as a philosophy of management dedicated to eliminating waste. The JIT philosophy strives for simplicity, minimal lead time, and value-adding functions throughout operations. While the basic philosophy can be simply stated, the changes necessary to achieve the objectives are often profound. JIT often results in implementation of process flow redesign, synchronization of production to demand, small lot sizes, and pull system scheduling.

The cumulative impact of these activities changes the nature of doing business and requires complementary changes in accounting, control, and performance measurement. Efficiency and detailed work center cost measurement—the underpinnings of many existing cost accounting systems—are unnecessary in JIT-based systems. Accordingly, cost accounting systems are being redesigned as more and more manufacturers adopt the JIT philosophy.

Pursuing Continuous Improvement

A primary objective of JIT manufacturing is to reduce manufacturing lead time. This goal requires elimination of wait time and move time. Reducing lead time enables a company to break the vicious cycle in manufacturing. In traditional environments, long lead times result in increased confusion and erratic performance on the plant floor. By reducing lead times, the entire process is simplified, and improved control is achieved.

Essentially, JIT keeps materials moving. Idle inventory is not value-adding. Productive inventory management is ideally a continuous flow process without stopping and storage. This smooth flow can be accomplished by establishing manufacturing cells that minimize between-operations movement and maximize processing efficiency.

Uniform plant loading, or linear production, is also a goal, as well as a result, of successful JIT shops. Because the firm makes what it needs, and no more, production can be balanced. Quality improvement is a prerequisite for uniform plant load. Consistent production exactly matched to demand reduces the danger of short shipments and lost sales.

An overriding theme in a JIT environment is continuous improvement and change. The goal is to simplify and to do it right. It involves all employees in controlling their destiny, pushing control to the lowest level. Yet, it does not suggest that high technology and robotics are the only answers to manufacturing problems. Mastering JIT manufacturing is one of the best ways a firm can prepare for extensive automation. It yields a flexible, balanced, and yet simple production process, the key to successful automation. As noted by Bill Wheeler of Coopers and Lybrand:

> A company that moves to an advanced manufacturing technology or automation environment, or subsequently, to CIM without going to JIT first to gain the simplicity, and to gain the visibility necessary to determine where the biggest bang for the buck in productivity is through automation is equivalent to putting whipped cream on garbage.[4]

The Strategic Use of JIT Manufacturing

JIT manufacturing can improve a firm's cost, quality, and responsiveness. Each of these areas link production to the stra-

---

[4]Interview with Bill Wheeler, 1986.

tegic position of the firm. Susman and Dean argue that the onslaught of advanced manufacturing techniques is not only affecting the manufacturing function in organizations, but is also leading to changes in the basic manufacturing strategies.[5] They suggest that single-criterion strategies, such as low-cost producer, are being replaced by low-cost/high-quality approaches that provide the customer with a superior product at competitive prices. JIT manufacturing supports this move toward complex strategies by reducing the cost of variety with simple, reliable processes.

Figure 2–2 captures the JIT cycle of success. Through the themes of visibility, synchronization, simplicity, continuous production, and holistic manufacturing, JIT affects each area

FIGURE 2–2
JIT Cycle of Success

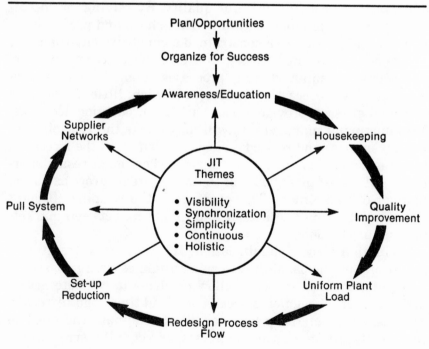

[5]G. I. Susman and J. W. Dean, Jr., "Strategic Use of Advanced Manufacturing Technology in the Emerging Competitive Environment," Working Paper, The Pennsylvania State University [CIMTOC], 1987.

of the organization. Quality enhancements evolve as linearity is pursued. Because quality products are easy to sell, they provide manufacturing firms with a strong niche in the market. Ultimately, costs are minimized as well, because value-added criteria and improved quality result in reduced rework, inspection, and warranty costs.

Low-cost/high-quality products are a salesperson's nirvana and a consumer's dream. When these products combine with the responsiveness inherent in a JIT process, there is little question that a competitive advantage is gained by the successful adoption of JIT manufacturing.

JIT Manufacturing: A Summary

In the traditional manufacturing environment, marketing strategy targets a specific market segment, where competitive advantage is based on price or quality. By relying on one dimension of the product to gain market share and profits, companies must operate in a constrained competitive environment.

These constrained strategies are likewise accompanied by unresponsive manufacturing processes. Long lead times require complex forecasting techniques which, little better than educated guesses, provide a poor basis for production decisions. Manufacturing managers, unwilling to bear the risk of obsolete inventories and related costs, often disregard the forecasts, building what their intuition suggests. The usual result is error on all levels: inventories of the wrong items grow, as do the number of back orders. This scenario is the jumping off point for manufacturing inefficiency and infighting between marketing and production.

Through process simplification, JIT solves most traditional marketing problems. Multicriterion strategies emerge based on flexibility, quality, and cost, allowing firms to compete across multiple market segments. Decreased lead times and increased responsiveness at the production level eliminate the need to rely solely on marketing forecasts or, equally error-prone, manufacturing's intuition. The result: manufacturing becomes a strategic weapon, differentiating a company and its product from competitors.

CRITERIA FOR MANUFACTURING EXCELLENCE

All other things being equal, the customer will buy the product that gives the best overall value. This is an integration of many factors—product features, quality, reliability, price, customer service, etc. This has always been the case, and it always will be.[6]

Traditional manufacturing strategy focuses on only one element of the product in developing and maintaining a market niche (e.g., low-cost producer). The Japanese are changing this approach to manufacturing strategy. They now compete on cost and quality, not only because of implications in the marketplace, but also because they know cost and quality are not mutually exclusive goals, but are reinforcing in nature.

A desire for manufacturing excellence is driving the technology movement in American manufacturing. The strategic

FIGURE 2–3
Migration Path of Technologies

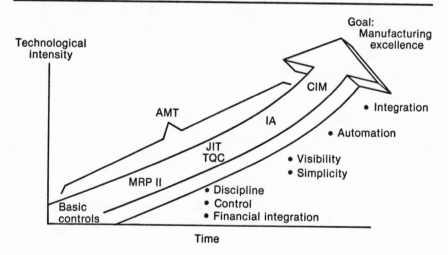

---

[6]From a speech delivered to the Association for Manufacturing Excellence on September 11, 1986, by William J. Weisz, Vice Chairman of the Board and CEO, Motorola, Inc.

goals subsumed by a desire for excellence include high quality, low cost, and high customer responsiveness. In other words, excellence in manufacturing entails strong performance on more than one product attribute, which is achieved by providing the customer with a final product that surpasses all others in value and reliability. Manufacturing excellence translates to the use of manufacturing as a strategic weapon.

Four primary criteria for manufacturing excellence are: (1) people, (2) quality, (3) delivery, and (4) cost. Achieving excellence with each of these elements will depend, to some degree, on the successful adoption of manufacturing and accounting innovations. As Figure 2–3 suggests, the final goal—gaining a competitive edge—results from accepting and pursuing those technologies, processes, and procedures that meet these criteria for excellence.

CHAPTER 3

TECHNOLOGY ADOPTION: LESSONS FROM THE FIRING LINE

Companies tend to adopt the new technologies in a piecemeal fashion, matching solutions to problems as they occur. This fragmentary approach to modernization makes the accountant's task all the more complex. Because each implementation is unique, it is essential that the new technologies are understood before they are translated into management accounting system changes. This chapter details the technological status of the firms included in this study, providing a basis for analyzing their accounting modifications.

DIESEL SYSTEMS, INC.: ADOPTING JIT IN A DECLINING MARKET

Diesel Systems' management is taking three major actions to meet its competitive demands: (1) automating only what needs to be automated (e.g., what makes sense); (2) dedicating itself to continual change; and (3) using all available resources, whether in accounting, manufacturing, or management, to enhance its competitive advantage. According to Richard Schonberger, world-class manufacturing "depends on blended management . . . rather than domination by a separate group of managers . . . which marshalls resources for continual rapid improvement . . . in terms of making maximum use of people and current machinery."[1] Company A reflects this approach to manufacturing.

MRP: Backdrop to JIT

The MRP system exists in relative obscurity at Diesel Systems, serving as a backdrop to current automation and advanced technology efforts. Implementing this system has reduced the number of expediters drastically: from 54 in 1984 to 0 in 1986.

Only the schedulers view and/or make changes to the MRP system. In scheduling, products are dealt with as families geared toward particular cells on the floor. Two factors drive actual line schedules: lead time for the finished product and a policy that limits the amount of inventory allowed on the floor at any one time.

Statistical Process Control

Quality is emphasized at Diesel Systems and is maintained on the floor by SPC techniques. Production employees maintain SPC charts and other visual aids. The charts indicate goals and progress toward those goals in terms of efficiency, lead/ inventory levels and times, and effectiveness. The quality enhancement obtained through this tool has helped Diesel Systems become one of the few American companies to achieve certification as a supplier to a Japanese manufacturer.

Management believes implementation of SPC procedures has provided the following benefits: improved overall product quality, detailed quality information for evaluating operator performance, reduction of scrap and rework, and elimination of quality control inspectors. Diesel Systems, through the use of SPC, has reduced scrap an average of 25 percent per year since 1984 (current level is $500,000). Quality control inspectors monitor operator-maintained SPC charts to ensure continued data and product integrity.

---

[1] R. Schonberger, *World Class Manufacturing* (New York: The Free Press, 1986).

Vendor Certification

A critical aspect of a JIT manufacturing process is the support and cooperation provided by a firm's vendors. Diesel Systems is pursuing a program of vendor selection and certification designed to reduce the total number of vendors (no duplicate sourcing) and establish certified vendor networks. To date, 8 of the 80 preferred suppliers have been certified and are no longer subject to incoming inspection. Overall vendor performance criteria are: percent of on-time deliveries, quality, and price competitiveness. While these criteria exist, management undertakes little formal monitoring of vendor performance.

JIT: Substantial Progress

In 1978, Diesel Systems committed itself to automation and the use of advanced manufacturing technologies. While this commitment continues, the plant can be described as a mix of technologies, running the gamut from laser and numerical control equipment to the traditional lathe and drill press operations.

Managers at Diesel Systems believe 80 percent of the savings resulting from JIT methodologies and techniques can be reaped simply by creating work cells. Their belief is that reorganizing the plant to JIT cells is directly responsible for reducing inventories from the 1984 level of $37 million to a current $15 million in investment. Relatedly, inventory floor space reductions exceed 19,000 square feet (28,000 square feet to a current level of 9,000). As noted by one manager, "Under JIT, the entire process is under better control."

With this in mind, plant management is identifying and creating JIT work cells. Sixty percent of the plant has been reorganized into JIT cells dedicated to part or product families, while 40 percent remains functionally oriented. The cells range from highly sophisticated robotic lines employing Kanban squares, electronic sensing devices, and various forms of in-line quality control to others created simply by placing old machines together.

Once cells are created, top management relies on production personnel to reorganize processes and achieve further efficiency gains in the production of components and subassemblies. Using end-unit demand levels as the trigger, components and subassemblies are pulled through production cells. Hence, the basic characteristics of JIT production are in place, even though some functional production continues.

Effective Use of JIT Techniques

Diesel Systems employs a JIT management philosophy, uses basic technologies such as MRP and SPC in its process control efforts, and is adopting robotics and flexible manufacturing processes on an as-needed, or process-justified, basis. Management is dedicated to continual improvement, as suggested by this final comment:

> The beauty of the cell approach is simplicity. Once we have a cell, we don't need accountants any more to tell us what is going on. . . . Once the cells are in place, we can push for decreased lead times by making supervisors responsible for picking and implementing opportunities to improve efficiency and effectiveness on the floor. What we are really trying to do is push the responsibility for controlling the process down to the lowest levels, where they can really effect change.

TELECOMM CORPORATION: STRUGGLING WITH HISTORY

The TeleComm Corporation plant studied produces personal computer components, such as printers, keyboards, display terminals, and controllers. The plant occupies 695,000 square feet and employs 1,300 direct laborers and 350 expense, or indirect, individuals. Production at this facility in 1986 was approximately $288 million.

Product Flow

Products assembled at the plant followed a zigzag path typical of most major U.S. production facilities today. Such a path results in an overwhelming percentage of non value-added (e.g., move and wait) time in the total production cycle. In a three-week average cycle time, 99 percent of this time is spent waiting and 1 percent is actual productive (i.e., assembly) processing, individuals at the plant estimated. Out of the 1½ hours of labor charged to products, it is estimated that only one third of this is tied to value-added functions. The process is so complex in the traditional assembly area that one manager noted, "If you put raw materials in, eventually it will belch out as product at the end."

MRP—Alive but Not Integrated

Although the facility has an active MPS/MRP system, it is neither mature nor integrated. Because of long lead times, the plant fails to hold a firm window on its production scheduling. The long lead times create two problems: (1) long planning horizons are dictated by the process cycle times; and (2) these long planning horizons result in frequent changes to the schedule. As suggested by one of the managers:

> We get caught up in being a corporation. We try to set up situations where we won't change the MPS/MRP system out one quarter, but then reality intercedes. Knee-jerk reactions to business cycles prevent us from solidifying our schedules.

The same manager noted there have been up to 15 major schedule changes in one quarter.

Inventories: Progress Being Made

The TeleComm Corporation plant observed during this research maintains in excess of $101 million in WIP inventory and $24 million in finished goods. Inventory reductions in 1986 were targeted to bring this number down to $91.5 million. Inventory turns currently number about 3½ per year. Policies geared to cut WIP levels have pushed these inventories down to 25 percent of their prior size.

Vendor Certification: Room for Improvement

TeleComm is just beginning to use vendor certification. To date, out of 3,000–4,000 total vendors, only 20 are certified. As currently practiced, the rigorous certification has been viewed as an expensive undertaking and hence is being pursued slowly. Five main stages exist in TeleComm's certification process:

1. Lot-by-lot examination by TeleComm Corporation quality control inspectors on-site at the vendor.
2. Subsequent lots inspected on arrival at plant on a sample basis to determine if quality levels set in step one are being maintained.
3. Periodic review/testing of incoming lots replaces total lot sampling approach.
4. Surveillance on yearly basis to see if approved procedures are still in place at vendor.
5. Total certification with no inspection.

This process, as well as most purchasing functions at Tele-Comm Corporation, is done on a centralized basis at corporate headquarters. Certification, therefore, is a purchased commodity for the individual plants, which partially explains its cost and slow adoption rate.

Islands of Automation: Successes and Failures

This site employs two predominant types of automation: high-speed CNC machines for the surface mounting of computer and printer cards, and robotics. The high-speed insertion machines operate with less than a 20 parts per million error rate. Of these errors, 99 percent are failures of purchased components. SPC is unnecessary in a process tightly controlled via technology. Hence, although further controls could be implemented, the individuals involved believe the associated costs would exceed the benefits.

Robotics have not been as successfully implemented in the plant. In one area, a pick-and-place robot moves boxes and loads storage bins. This has proven very beneficial, in that a large number of physical injuries occurred in this area before a robot was installed.

An area of automation that has not resulted in the same savings is keyboard assembly. It is the first "focused factory" attempt at this facility, and has not achieved its full potential. Robotics were installed to perform the assembly operations.

Whereas initial estimates indicated 3,000 keyboards would be produced per day, production at this facility is currently less than 800 units a day. The result is idle robots used to only 20 percent of their planned capacity. As noted by one manager:

> We departed from the standard rule of only using robots in jobs where factory workers couldn't do the work because it was dangerous or boring. Now it looks like a bad idea as we're not flexible given the fact that plans have changed and technology has changed. People are more flexible than robots . . . people will change the job to make it better. I have yet to see a robot beat a person who enjoys doing a job.

It appears that robotics are *not* the answer to every production problem.

Computer Assembly and Test

The computer assembly and test (CAT) manufacturing facility was the focal point of study at TeleComm Corporation. The CAT facility is a total JIT shop consisting of a continuous flow of materials from the loading dock back to the loading dock with almost no non value-added handling.

The JIT shop is a de novo production process given free rein by the corporation to design its own facility and performance measurement system. Using a single document production tracking system, information processing has been revamped. Accounting systems are separate from established corporate systems. An initial agreement minimizes plant-based allocations to the area; rent is charged on a square footage basis to cover basic services and costs. Utilizing 20,000 square feet of floor space, the JIT shop can produce more than $80 million a year of product.

The process itself is designed to flow dock to dock in less than four hours. The assembly process is broken into nine major operations, and assembly time is divided into "beats"— 2½-minute intervals used to synchronize the process. Kanban squares and racks control the production flow and on-floor inventory levels. Out of the total cycle time, 34 percent is actual production (value-added), while the remaining time is spent in queue and move. Contrasting this to the three-week lead time in the plant underscores the potential savings and increased responsiveness of the JIT shop.

COMPSCI INDUSTRIES: ORGANIZATION AROUND LINKAGES

In 1984, CompSci Industries shifted from a focus on technology to a marketing basis. Before 1984, the company was organized around its basic technologies: electronic instruments and computers. Today, it focuses on establishing lines based on market niches and demands. This shift in strategies is reflected in structural changes in the organization, such as the elevation of marketing management to the level of executive vice president, as well as a reorientation of its production and selling areas.

Simplified Processes

One striking feature of CompSci Industries' plants is their clean, simple, and extremely quiet production processes. Noise is minimized or contained through separation where necessary. Only the flow of materials from the receiving dock to finished goods signals the existence of a highly productive assembly process. CompSci Industries is rapidly approaching the status of total JIT assembly and is utilizing Kanban squares and focused work cells to move the responsiveness of the system forward, decrease lead times to their minimum (e.g., eliminate non value-added time), and achieve linearity in production.

CompSci Industries has a mature MRP II system, a mature JIT manufacturing process, is utilizing only those technologies that fit the simple production processes used in the plants (e.g., CNC machines), and is not contemplating CIM, mainly because it is deemed to be both unnecessary and expensive. Pursuing a manufacturing strategy based on simplicity resulted in the following accomplishments:

- Cycle time has been reduced from 22 days to four hours.
- Overhead per unit has been reduced 30 percent.

- Quality has been improved more than three times the goal voiced by top management in half the anticipated time.
- Material costs are down 12 percent.
- Scrap has been reduced 30 percent.

Vendor Certification

CompSci's dedication to vendor certification reduced inspection and subsequent direct delivery of incoming materials to the production line. Eighty percent of the vendors used are certified. The remaining 20 percent fall into two categories—subcontracted assembly, such as fan assemblies jobbed out to handicapped cottage workers, and generic parts currently not single-sourced.

CompSci Industries rigorously pursues improved vendor relations through an outreach program. Suppliers of vital components are often involved in the design of new products. In this way, CompSci harvests their expertise on how to best utilize their technology and products. Actual vendor evaluation is an ongoing process that utilizes a feedback system that scores performance on delivery, price, quality, and lead time.

Reducing Lead Time

Lead time concerns continue to plague CompSci Industries. JIT is built on a goal of zero lead time. While such perfection is probably unattainable in a manufacturing system as geographically dispersed as that in the United States, constant reduction of lead time can be pursued. This translates into increased efficiency and improved production flows for the receiving firm. One spokesman at CompSci said the firm pays a premium for decreased lead time because its product gets to the market faster (e.g., increased responsiveness), not to mention the decreased cost that follows due to reductions in safety stocks and, hence, inventory carrying charges.

This philosophy is not applied indiscriminately to the components and raw materials CompSci uses. The company utilizes a two-tiered raw material categorization process that separates these items into A and C groups, rather than focusing on A-B-C items, as is traditionally done.

"A" items represent 10 percent of the total components but 90 percent of the total material cost for a specific product. For these items, raw-in-process (RIP) inventory buffers are ½ to 2½ weeks, with an average of 2 weeks. In contrast, C items (90 percent of part count, 10 percent of value) are buffered from two weeks to three months, based on contract terms and relative dollar value. Additionally, new materials are screened to avoid duplication of part numbers for the same component as well as to allow for corporate group purchases where feasible.

Contracts—Not Orders

In another vendor policy, CompSci Industries replaces individual orders with contracts. Contrary to an evolving practice followed by many companies that issue open orders for specific parts, this firm wants suppliers that commit to their relationship with the company. Instead of standing orders, rolling contracts spanning approximately eight weeks of production are used. This supports the vendor evaluation process while maintaining an alternative to sole-sourcing. Although some exceptions are made for unique raw materials, this policy is generally applied to all vendors.

Results: Decreased Cycle Time

These inventory control policies allow CompSci Industries to focus on decreasing production cycle times. Queue time has virtually been eliminated. Actual production time (all value-added) is approximately 45 minutes, down from four hours several years ago. Total time from order of materials to finished goods shipment is less than three weeks (e.g., roughly 2½ weeks in RIP, 45 minutes on line, followed by one to four days in finished goods). The firm believes there is still room for improvement and is seeking ways to reduce RIP and finished goods wait time. Decreased cycle time effectively increases capacity of a plant, a JIT side benefit often overlooked in the literature.

These improvements have given CompSci Industries a competitive edge based on 33 percent cost of goods sold, 11 to 12 inventory turns per year, and an asset turnover ratio of 3.85. The ultimate goal, however, is to meet customer needs. As noted by one manager:

> The ultimate scorecard is your success in the market. It tells you what game you are playing; after that you can decide where you want to be in that game. Increasing volumes gives you the room to maneuver, to plan rather than react. The way to get there, then, is to measure what is important to your success in the market, not what you did in the plant three weeks ago.

At CompSci Industries, this approach resulted in a cost of goods ratio of 4 to 1 (80 percent materials and only 20 percent overhead).

MICROCHIP, INC.: RENEWAL AS THE DRIVING FORCE

MicroChip, Inc., is dedicated to growth through expansion of its markets, coupled with renewal of its manufacturing efficiency. The company is focusing on remaining viable in its established markets. It seeks to achieve this long-term growth by constantly renewing its manufacturing efficiency, by striving for continual quality improvements, and by placing its managers in an on-going training program that emphasizes manufacturing strategy and systems needed to maintain a competitive edge. This process pursues low costs, high quality, on-time delivery, and flexibility in its products and processes.

MicroChip broke into the Japanese market in 1986 by pursuing these objectives. Besides marketing directly to Japan, the company also established a plant in Japan: a jointly owned semiconductor manufacturing facility. The firm sees this alliance as a way to both enhance the flow of technology and gain increased access to the Japanese market. This view reflects a belief voiced by the company's chief executive officer: "When

we say 'win,' we mean being successful, standing toe-to-toe
with any competitor. Both parties must win—remain viable—
remain strong." Rather than pursuing protectionism, Micro-
Chip is eager to square off in a competitive battle to make the
best products.

A Mature JIT Process

The MicroChip facility used in this study produces applica-
tions-specific integrated circuits (denoted ASIC). Although
ASIC employs a simplified, mature JIT manufacturing process,
it is faced with a complex product mix. Its main product is a
custom logic board on a chip. One of 15 fabrication shops run
by MicroChip, ASIC produces approximately 1,500 units per
day in a 1,000-square-foot facility.

ASIC undertook the move to JIT techniques and processes
in 1983–84. There was tremendous room for improvement with
29 cost blocks and almost a two-month assembly and test cycle
time. The 29 cost blocks were reduced to 11 JIT cells, which
simplified data collection and streamlined production. The final goal at this site is to have only four cells: assembly, burn-
in, heat sink, and test.

Cycle Time Reductions

These moves have translated into dramatic cycle time savings.
Assembly time, formerly taking 21 days, is down to 3 days. Test
processes, which used to take 25 to 30 days to complete, are
completed in 5 to 6 days. The existing cycle time target for the
entire process is seven to eight days. The goal, which will be
reached through further simplification of the process flow, is to
reduce total cycle time to two to three days.

Delivery cycle time is also being addressed through the
overall focus on process improvements at ASIC. While it currently takes six to seven weeks from receipt of a customer order until it is delivered, plans are to reduce this time to two to
three weeks. Raw materials procurement constraints represent
the difference between the two to three days needed to actually
produce the order and the days projected for total cycle time.

This appears to be an optimal solution for ASIC, which faces a complex, customized product mix. The only remaining question is whether standardization and simplification of the products themselves (e.g., the principles of design for manufacturability) can further enhance the responsiveness of the firm.

Inventory Reductions

After implementing JIT on the plant floor, inventories at ASIC decreased from $600,000 to $200,000. The following statement reflects MicroChip's attitude toward inventory:

> Regardless of the reasons that material is in the plant, it represents money tied up. The more money tied up, the lower the inventory returns and lower return on net assets.

Bootstrapping Performance Measures

At MicroChip, Inc., cycle time reduction is the key issue. Given the inherent relationship of inventory levels to cycle time, management examines inventory levels by cell and, using the following formula, determines cycle time:

$$\frac{\text{Cycle time}}{\text{in weeks}} = \frac{\text{Average inventory}}{\text{Weekly output}}$$

This formula is applied to segments of the inventory to pinpoint areas for improvement and to track the impact of process changes for final savings.

Quality and Yield

Quality and scrap levels are reported twice daily to plant management at ASIC. Current yields on wafers approach 97 percent, while the probability of producing a good die bond for the integrated circuit is only 80 percent. Die bond problems plague ASIC production, affecting linearity, quality, and scrap levels. The actual fabrication process blocks further improvements.

MicroChip has not fully adopted SPC, which is currently used primarily as a maintenance tool for guiding preventative repair schedules. Although one company spokesperson noted that engineering and design should be the areas in charge of SPC and its proper use in the plant, the company has been slow to recognize the value of statistical techniques for improving quality performance.

GENERAL BUSINESS, INC.: CIM AND A GLOBAL STRATEGY

Two key aspects of *asset utilization* are targeted for development and enhancement in General Business's total manufacturing strategy: automation and inventory. The focus is on designing for automation to reap quality improvements while reducing overall manufacturing costs.

Reducing part numbers is one primary area of concern for General Business. Total parts are perceived to be a major cost driver in assembly as well as related to product failures and inversely tied to quality and reliability of the product.

General Business is studying four elements of inventory management: turnover management, schedule stabilization, reduction of order churn, and pipeline management. These aspects of production are under fire from two sides: one stresses increased flexibility and the other, improved marketing/distribution channels. Better anticipation of market demands is one of the primary objectives stated in the corporate goals. A marketing orientation entails creating a product suited to customer needs rather than to production characteristics. Coupled with flexible, responsive production processes, the total impact on inventory management is dramatic. During the site visit, several people noted their performance measures were being increasingly oriented toward effective, efficient inventory management.

On the *quality* front, the manufacturing strategy is to achieve product quality leadership, surpassing prior company products as well as competitive products. The emphasis is on maintaining quality at each site. The ultimate objective of zero defects is being pursued through stress screening of products before shipment, failure analysis, and vendor awareness to provide feedback and support process improvements.

The bottom line in General Business's manufacturing strategy is keeping these plans in line with overall business goals, goals built around responsiveness to market demands (e.g., availability, cost, and quality) and efficient use of assets. These manufacturing strategies, reflected in the CIM facility visited during this research, are detailed in the next section.

A CIM Environment

The General Business site is a fully automated, computer-integrated assembly plant producing laptop computers for the domestic and international market. Having reached maturity in its MRP II system and JIT processing capabilities, the firm has moved into other technologies on a case-by-case basis, focusing on efficiency/flexibility trade-offs as well as future strategic concerns in specific markets.

The CIM plant has two production lines. Utilizing 13 robots and no direct labor, each line can turn out a computer every two minutes. A single workstation is able to perform up to 12 separate steps, ranging from actual assembly of the laptop to placing a part in position for the next station.[2] This translates to a limit of 156 possible steps in the entire production process.

Design for Manufacturability: A Key Element

This constraint requires design for manufacturability and simplicity in the design of parts and packaging. The complexity resides in the robots rather than the process, an approach which suggests that General Business has learned the lessons of simplicity embedded in JIT. Using simple methods, well-designed products with a minimum of parts, and special component packaging to meet the constraints of the robots' capabilities, General Business reduced product costs while improving quality, reliability, and responsiveness overall. Coupled

---

[2]At General Business, Inc., the researchers were provided with a detailed article by B. Saporito that served as the basis for this exposition of the CIM plant visited.

with this radical change in product design, General Business has adopted an evolutionary product strategy to replace revolutionary designs of new product families. This strategy relies on standardization of technologies, parts, and subassemblies across the various families of products.

Statistical process control has been adopted throughout General Business and is being pushed back through the vendor chain to ensure high quality as well as low inventory-generating production of various items. The production precision necessary to keep a CIM plant running requires careful planning and control of component specifications. Even smart robots cannot sort bad units from good—consistency and quality are critical. Lack of these characteristics translates to downtime and significant cost problems.

MIGRATION THROUGH TECHNOLOGIES

The pursuit of manufacturing excellence is like climbing a path—each step brings new technologies and new challenges.[3] While MRP II teaches discipline, control, and financial integration, JIT focuses on simplicity and visibility. Processes under control in a JIT environment can be effectively automated. The final goal, CIM, is an integrated approach to manufacturing based on robotics and computerized production processes, designed to transform manufacturing into a strategic weapon.

A Migration Path of Technologies

One way to rank the companies, or the plants, observed during this research is on the basis of the technologies they have in place. This ranking, for the most part, is site specific, because the plants visited were selected based on their use of advanced manufacturing technologies.

---

[3]It is currently believed that leapfrogging technologies does not maximize a firm's potential benefits from or movement toward excellence because basic lessons are embedded in each technology.

TABLE 3–1
Technology Grid

Company	MRP II	JIT	Islands of Automation	CIM
Diesel Systems, Inc.	Integrated, mature	Mature	Negligible	Not applicable
TeleComm Corporation	Not integrated	Pilot	Robotics; auto-insert	Not applicable
CompSci Industries	Essential (evolved)	Advanced	Not relevant	Not relevant
MicroChip Inc.	Immature	Advanced pilot	Not applicable	Not applicable
General Business, Inc.	Mature	Mature	Not relevant	Mature

Table 3–1 provides a technology-by-technology analysis of the sites studied. While General Business is obviously the technological leader among this group, having several mature CIM plants, a question remains as to the relevance of different technologies.

If the sites observed during this research were placed on the technology migration path based on the characteristics detailed in the associated table, General Business would once again appear a clear leader on the path to excellence.

AMT Implementation: A Strategic Issue

The assumption inherent in both of these rankings is that one technology (e.g., CIM) is superior to another (e.g., JIT). Most recent research efforts and articles looking at the U.S. manufacturing environment have made this assumption, although it is seldom cited specifically. In fact, if CIM were to be implemented without mastery of the short-cycle management lessons inherent in JIT (e.g., CIM with batches and buffers), it could prove to be worse, in terms of efficiency and effectiveness, than JIT-based manufacturing. The sample used here illustrates the possible mixing of strategic decisions with technology applicability.

What may be evolving in practice is a phenomenon not yet clearly understood in the academic or popular literature—that different technologies are more applicable to certain product/ marketing strategies than others. Reviewing the discussion of CompSci Industries versus General Business, Inc., the strategic implications of their technology decisions become quite relevant.

Assume, as is popularly believed, these two firms are proactively managed, strategically oriented organizations competing in a rapidly changing, and hence uncertain, environment. Both companies have learned the lessons of JIT, namely simplicity and visibility. CompSci Industries has opted to remain a JIT shop, while General Business is implementing CIM. Does CompSci rank below General Business? On what criterion?

Although the identities of these two firms cannot be revealed, those who know them would more likely perceive them as merely different rather than as more or less successful in their move toward manufacturing excellence. Both have opted for flexible products built on a basis of evolution rather than revolution in design and function. Both are known for their strong cultures. One has chosen a complex technology utilizing robots, the other a simple technology using people. The relative merit of these two strategies is a question to be answered only by hindsight. In contrast to the implied rankings used to build the migration path described here, the actual findings of this study suggest it is the match of technologies to strategy that may be the critical linchpin to achieving manufacturing excellence.

CHAPTER 4

MANAGEMENT ACCOUNTING: BARRIER TO CHANGE?

The field of management accounting is in crisis. Traditional measurements and concepts are being challenged as the management accountant's role in the organization is questioned. Cost accounting is being blamed for the decline in productivity in the United States, as well as for promoting a mentality of managing by the numbers. For practicing management accountants, it is a time of turmoil—and of opportunity.

MANAGEMENT ACCOUNTING: A HISTORICAL PERSPECTIVE

The roots of management accounting can be traced to the origins of managed enterprise, about the first or second decade of the 19th century. Under the umbrella "cost management," cost information was used to evaluate the efficiency and effectiveness of resource consumption by manufacturing companies. This is in contrast to cost accounting, "the practice of attaching factory costs to units of product to value inventory and cost of goods sold."[1]

The internalization of various exchange transactions (e.g., dedicated labor force versus contract labor) by 19th century en-

---

[1] H. Thomas Johnson, "The Decline of Cost Management: A Reinterpretation of 20th-Century Cost Accounting History," *Journal of Cost Management*, Spring 1987, p. 6.

trepreneurs provided the impetus for cost management systems. Contrary to prior findings, the development of cost management systems preceded, or occurred concurrently with, the Industrial Revolution. Without this tool for evaluating the relative efficiency of operations performed within the firm against market solutions, it is difficult to conceive of the rapid growth of large, managed enterprises.[2]

The Era of Scientific Management

Turning the clock forward to the late 19th and early 20th centuries, the next major force in the development of management accounting can be traced to the Scientific Management movement. This cadre of engineer-managers developed a set of predetermined standards to evaluate labor and material efficiency, replacing the market prices used by earlier enterprises. It is important to remember that the environment was different then. External reporting was limited and unregulated and therefore had little or no impact on the development of information for management. Early standard costing systems were used for three primary purposes: (1) to gauge the potential efficiency of tasks or processes, (2) to differentiate between variances due to controllable conditions and variances caused by conditions beyond management's control, and (3) to simplify the task of inventory valuation.[3] The first two uses evolved from a need to control and assess the efficiency of the production process. The last use came from a recognition by financial accountants that this information could simplify inventory valuation for use in published financial statements.

Whatever Happened to A. Hamilton Church?

A. Hamilton Church, and his devotion to traceability, has been recently rediscovered by accounting historians. Church was a contemporary of Frederick Taylor and a major figure in the Scientific Management movement. While Taylor and his counter-

---

[2]Ibid., pp. 50–51.
[3]Ibid.

parts turned their attention to developing standard costing systems, Church explored and refined a method for trading costs via the "machine-hour labor rate method." In discussing Church's approach, Dr. Richard Vangermeersch, of the University of Rhode Island, writes:

> Now, Church attacked what he called his "friend," overhead. We regard overhead as an enemy. Why? Because we regard overhead as a glob. But Church said, "Get rid of that glob." And if you can take overhead, and break it into multi-factors of production, you are going to answer most of the problems of accounting and technology; because you can then control each of them. The most poignant one is machine labor hour rate method, in which you collect the cost of a machine and assign it as you would a direct laborer's cost.[4]

Church maintained that accountants and managers were in error in their single-minded attention to determining direct and indirect costs on a product basis. He believed the real source of cost was the underlying process and these processes should serve as the basis for a chargeback system to the individual products. This reflected his belief that one should consider the consumption of resources by various products when analyzing the profitability of both the products and the organization.

Many of Church's ideas appear to provide answers to current problems in the MAS. The question is, What happened to A. Hamilton Church? Why were his ideas lost in history, only to be rediscovered by able, dedicated accounting historians at a time of crisis? Church's dedication to tracing indirect costs to their cause required intensive data collection and analysis. The cost of this form of information was difficult to justify in a manually maintained MAS. Church utilized this model in his own business, but his methods never gained the widespread acceptance of the less complex standard cost models of his counterparts. A man born before his time, Church was, until recently, a lost figure in the history of cost management.

---

[4]R. Vangermeersch, "Milestones in the History of Management Accounting," *Cost Accounting for the 1990s, Proceedings, The National Association of Accountants, 1986,* p. 78.

Losing Our Identity: A Recent Phenomenon

In many respects, the fate suffered by Church mirrors the basis for the demise of cost management systems. Information-processing constraints made it increasingly important to seek efficient, low-cost, and flexible (e.g., for managerial and financial reporting) means of maintaining product cost information. H. Thomas Johnson notes, "Cost accounting was invented around 1900 by auditors whose concern was financial reporting, not cost management."[5] The need to value inventory without adding nontransaction-based data to the general ledger limited the types of information on which the auditor could rely. In a setting where generating information was very costly, the numbers produced by the auditor proved to be a low-cost alternative to the burdensome internal techniques employed to cost products.

Although many contend that the demise of cost management was independent of the development of audited financial statements, they were in fact related. The mandated use of cost accounting for financial statements gave it a leg up on cost management. Therefore, it appears information-processing costs may have led to the demise of cost management.

Academia: Out of Touch

Why did academics fail to recognize the growing obsolescence of existing cost accounting procedures? This question will be discussed at length, but it is unlikely that an answer will ever be found. Perhaps the economists' adoption of the simplified model of the firm prompted this myopia. Alternatively, the quest to add rigor to academic accounting research may have dimmed the attraction of complex, real-world problems that did not yield easily to the scientific approach.

The shortcomings of academia translated into gaps in the training of new management accountants. In fact, management accounting became the poor cousin to financial account-

---

[5]Johnson, p. 7.

ing education. The history and richness of cost management was lost, replaced through neglect by cost accounting and a financial orientation. In the long run, the current crisis may prove to be a blessing in disguise. The attention being brought to bear on management accounting procedures is already yielding solutions to the complex problem of updating an obsolete system.

COST MANAGEMENT ISSUES: CURRENT STATUS

It is becoming increasingly obvious that an MAS designed to support the demands of financial statement preparation is an inadequate source of information for management decision making. So why has the arbitrary, financially oriented approach dominated? A brief review of early management accounting texts suggests the weaknesses of the current system can be traced to inaccurate application of underlying principles, rather than to inherent flaws in management accounting. Management accountants have failed to stick to their knitting.

Cost elements, once critical for effective cost management, are now less important. A concern for quality, timeliness of delivery, the reduction of fixed costs on a value-added basis, an enhanced system for tracking development costs, and recognition of product life cycles need to be incorporated in the reporting system. The following discussion elaborates on these issues.

Changing Cost Patterns and the MAS

Two major types of changes in cost patterns are associated with AMTs, as depicted in Figure 4–1. First, the basic components of product cost change as direct labor becomes an increasingly insignificant part of product cost. Relatedly, equipment and technology costs grow as manufacturing facilities focus on automation. Not only are more functions being performed mechanically, but also the technological life of these assets is decreasing, creating a large pool of costs that must be

FIGURE 4–1
Cost Pattern Changes

Changing Cost Behavior Patterns

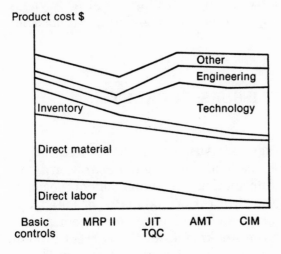

Product cost $

Existing Accounting Emphasis

Accounting effort

OH	10%
Material	10%
Direct labor	80%

Time

quickly recovered. Finally, as the use of computers increases, the fixed component of information-processing cost grows. To date, little is known about the value-added characteristics of this information technology.

Indirect Costs Dominate

The current literature maintains that indirect costs are replacing direct costs in the total cost package because of the substitution of support labor for direct labor, machines for people, and computers for clerks. Direct costs are traceable to individual products or processes, while indirect costs are less closely tied to specific operations.

As the number of traditionally defined, indirect costs increases, development of new allocation bases is critically important. Many of the traditional assumptions that provide the basis for overhead allocations, variance analysis, and the other basic tools of the management accountant are no longer valid.

Distorted Product Codes

Cooper and Kaplan[6] suggest the complexity of production rather than its volume drives many overhead costs. Historically, the use of labor hours proxied this relationship, because more man hours were needed to complete intricate tasks. This relationship no longer holds, as complex tasks are performed by machines maintained by people.

Therefore, another outcome of changing cost patterns is the distortion of product costs. The allocation of an increased pool of indirect costs over incorrect activity measures creates this distortion. The result is excessive burden rates and the shifting of cost among products. The existing allocations increasingly fail to match resource consumption with cost apportionment. The rapid increase in technology costs lies at the heart of this distortion, which is exacerbated by the use of direct labor—the *replaced* component of product cost—as the allocation basis. Figure 4–2 illustrates the impact of faulty allocations on product costs.

---

[6]R. Cooper and R. Kaplan, "How Cost Accounting Systematically Distorts Product Cost," working paper, Harvard University Graduate School of Business Administration, 1986.

FIGURE 4–2
The Impact of Allocations

Product A is technology-intensive. Product B is labor-intensive. Using a value-added approach, the true cost of both products:

	Product A	Product B
Direct labor	$ 50	$200
Direct material	300	300
Direct technology	200	50
Other overhead	375	375
Total product cost	$925	$925

The resulting overhead rate, based on direct labor dollars, would be computed as:

$$\text{Overhead} = \frac{\text{Technology + Other overhead}}{\text{Total direct labor costs}}$$
$$= \frac{(200 + 50 + 375 + 375)}{(50 + 200)}$$
$$= \frac{1000}{250} \text{ or } 400\%$$

The traditional cost-accounting approach would apply the overhead to the direct labor dollars using the 400% overhead rate. Using this approach, the reported product cost for A would be $550 and for B would be $1,300, a 40% error.

	Product A		Product B	
Direct labor	$ 50		$ 200	
Direct material	300		300	
Overhead	200	($50 × 400%)	800	($200 × 400%)
Total product cost	$550		$1,300	

Inventory Valuation

Inventory valuation issues are closely linked to product costing in the current environment. The goal of zero inventories inherent in AMTs highlights two major problems for inventory management. First, WIP carrying costs are obscured by various aspects of the existing MAS. Because of the financial orientation of cost roll-up procedures, recorded inventories are constrained to direct costs plus a fair portion of indirect costs.

While the costs of carrying inventory are one of the basic components of such models as the economic order quantity (EOQ), they are largely ignored in the accounting treatment. The fact that due to the conservatism principle, financial accounting is restricted to recording costs that are traceable to specific transactions does not fully explain this shortcoming. Once again, the inadequacy of MAS procedures appears to be tied to a failure to invoke the traceability criterion.

Second, inventory tracking and costing have consumed tremendous resources simply because of their magnitude. In a zero inventory setting, this function is unnecessary. Therefore, the advent of AMTs raises more than the issue of resource consumption. If the major task most management accounting departments perform is inventory valuation, the concept of zero inventory bodes ill for the profession. But if the MAS expands to fill its other neglected roles, such as planning support, this trend will prove to be a boon rather than a curse.

A Search for Timeliness and Relevance

Johnson and Kaplan[7] suggest that financial managers, relying exclusively on periodic financial statements for their view of the firm, are suffering from management myopia (e.g., a short-term focus that places the long-term viability of the organization at risk). In relying on arbitrary time frames, such as monthly reporting, for information, the management accountant supports the move away from strategic management to reactionary, short-term profit making.

Current management accounting reports may be neither timely nor relevant. Given the rapid change in today's manufacturing environment, reports generated after the fact are of little use. While getting the reports out faster may seem the simple solution to one problem, gathering the relevant numbers is not so easily resolved.

---

[7]H. T. Johnson and R. Kaplan, *Relevance Lost: The Rise and Fall of Management Accounting*, (Boston, MA: Harvard Business School Press, 1987).

A second point made by Johnson and Kaplan is that the nature of fixed costs is more variable than has been recognized.[8] They argue this distortion occurs from the monthly cutoff used to separate accounting periods. While many costs are sticky, responding slowly to changes in volume or complexity of production, they are not constant. Variability may be a third aspect of the change in cost patterns that dictates the need to change the MAS.

At this stage, the management accountant might shout, "Enough!!!" It is difficult to accept that the choice of a monthly reporting period could be responsible for management flaws. Yet management accountants have long been aware of the fact that people will focus their activities to maximize performance against the numbers tracked by the management reporting system. This is called goal displacement, or means-end inversion. Given an objective to support management decision making in the areas of planning, coordination, and control, some responsibility must be taken for neglecting environmental scanning and strategic planning.

MAS Shortcomings

Embedded in these discussions is a series of specific factors that lie at the heart of the current crisis in management accounting. The key issues are:

- The distinction between direct and variable cost, as well as indirect and fixed costs, has blurred.
- The focus has turned to the source of costs (e.g., drivers) from a preoccupation with variances and standard costs.
- Increased recognition of the interdependence between cost/performance among organizational subunits has negated the traditional focus on organizational cost control.

---

[8]Ibid.

- Change in manufacturing processes has shifted a significant portion of product cost from traditional direct cost to indirect, resulting in high burden rates that distort true product costs.
- New information-gathering devices and techniques have made cost traceability possible on a more detailed level.
- Compression of the life cycle has shortened the period available for recovery of development costs, necessitating efficient and effective production techniques from inception.
- Recognition of the cost of inventory is placing new emphasis on measuring and reducing cycle time.
- Focus on eliminating waste is leading to increased demand for value-added measurements of performance.

MAS WEAKNESSES: A TECHNOLOGY-BASED APPROACH

Kanban and Its Effects

Looking at the various technologies reveals a more specific list of shortcomings of the current MAS. For example, in a Kanban system, the work-order-based, production-unit focus of the traditional MAS needs to be deemphasized. It should be replaced by a form of process costing that reflects the goal of a continuous flow of materials through the plant. One main advantage of a Kanban system is this movement toward process manufacturing.

The nature of Kanban systems reduces the reliance on individual work orders and material requisitions to monitor WIP inventories and signal their movement through the plant. The visual simplicity of process movement encompassed in the use of Kanban squares significantly limits the amount of WIP inventory, reduces the number of accounting transactions needed to track inventory movements for efficiency analysis, and replaces a sluggish transactions-based monitoring system with a highly responsive process control mechanism.

NC Technologies and the MAS

The implications for the MAS change drastically in NC technology. David Keys detailed six MAS problems related to use of CNC machines:[9]

1. Identifying wages of machine operators as direct or indirect labor is difficult.
2. Direct labor is becoming increasingly fixed and less variable.
3. Benefits derived from NC machines are difficult to quantify.
4. The short run is overemphasized to the detriment of the long run.
5. Overhead allocations are not accurate and result in poor decisions.
6. Performance of NC machines is difficult to measure.

Some of the weaknesses attributed by Keys to the NC environment reflect general problems discussed earlier. However, the impact of NC technology on the relationship between direct and indirect labor adds new concerns to the list of MAS shortcomings existing in a Kanban system. Labor reporting revamping is added to the materials-related changes inherent in Kanban, placing additional stress on the outmoded MAS.

The MAS and Other AMT Approaches

Islands of automation (e.g., robotics) represent another step up the migration path. As noted by Allen Seed, III:

> An automated factory feeds work into the cost center, processes it with one or more machines, and removes it from the cost cen-

---

[9]D. Keys, "Six Problems in Accounting for NC Machines," *Management Accounting*, November 1986, p. 38.

ter without human intervention. One person may tend a system of a dozen or so machines, and this person tends to spend more time dealing with trouble than operating the machines. In such circumstances, burden rates based on direct labor may approach infinity, and production costs bear little relation to the amount of work the machine operator does.[10]

Taking these issues one step further, Dilts and Russell[11] note than an IA environment not only affects the relationship between direct and indirect labor, but also increases the demand for immediate feedback from the control system on the performance of the production process. IA technologies require a shift in allocation basis, a focus on controlling and eliminating overhead, and an increase in emphasis on technology costs. The current MAS can no longer function in such a drastically changed environment.

In a CIM environment, direct labor is removed from the product cost equation, as materials become the major component of product cost. Elimination of waste and inefficiency in materials-related transactions is one of the key measurements used to manage the CIM plant. Design for manufacturability is relied on to decrease scrap and rework and provide better utilization of facilities as total quality control is achieved at the start of a process or procedure. A CIM environment demands total performance measurement based on value-added criteria, focused on ensuring maximum efficiency and effectiveness at every level of the organization.

JIT and the MAS

Can JIT be implemented using today's costing measurement? It is very difficult to implement JIT because the tenets of efficiency reporting and traditional cost accounting approaches

---

[10]A. Seed, III, "Cost Accounting in the Age of Robotics," *Management Accounting,* October 1984, p. 40.

[11]D. M. Dilts and G. W. Russell, "Accounting for the Factory of the Future," *Management Accounting,* April 1985, pp. 34–40.

run counter to the JIT philosophy. While machine and labor efficiency ratings encourage production of large batches of inventory, which is supported by the need to absorb overhead costs, a JIT environment focuses on producing only what is needed. Inventory is eliminated because it causes costs that are not value-adding.

In a JIT-oriented manufacturing system, effectiveness (e.g., doing the right job) is given primary emphasis. Efficiency is defined as doing this job the best way possible. Efficiently performing the wrong task—such as building inventory to absorb overhead—is not only undesirable, but also cannot occur if a true JIT philosophy is guiding decision making.

JIT manufacturing moves the accounting system toward process costing techniques. With JIT, very little time exists between operations because of effective use of the cell concept and group technology. Additionally, there is effectively no inventory between operations (e.g., only what the Kanban square can hold), and high velocity of material movement eliminates the ability to track single components or units through the production process. Johansson of Coopers and Lybrand said of this process:

> JIT creates a very high-velocity level of output, with lead times that really defy the accountant's ability to track each piece moving through the process. So I see using a process cost accounting system. It looks process, feels process, and so it must be process.[12]

Process costing is easier to use, because costs are accumulated by the process or operation rather than by piece. It is also easier to maintain than traditional job-order costing systems. The cost of processing the part or product remains relatively unchanged unless a change occurs in the process itself. Also, it is cheaper to use because fewer resources are required for data accumulation, processing, and report generation. A movement toward the efficiencies inherent in a process costing

---

[12]H. Johansson, "The Effect of Zero Inventories on Cost," from a joint speech with T. Vollman and Vivian Wright, *Cost Accounting for the 1990s, Proceedings, The National Association of Accountants*, 1986, p. 146.

approach will simplify the MAS, decrease the cost of the system, and provide for more relevant and timely reporting.

TECHNOLOGY: THREAT OR OPPORTUNITY?

> The bad news is that much of what management accountants have done and are doing in America's factories will no longer be required in the factory of the future. The job will be passé. The good news is that those management accountants who can adjust to the changing factory environment will still hold jobs, and more important ones. . . . The objective of all of this is to make money. . . . The right combination of customers, productive capability and products will maximize your cash flows as well as the return on your invested capital. Your job as management accountants is to help your managers find that combination. I don't think you do it exclusively through the cost accounting system. I think you do it to a much greater extent by getting out into the factories, understanding what is going on operationally, and contributing to the operational management of your businesses.[13]

Does the factory of the future entail the demise of the management accounting profession? Is the profession doomed to go the way of the buggy whip? If you buy into what Robert Howell says above, the final outcome of the current wave of change in the manufacturing sector is up to the management accountant. Chained to the bread-and-butter tasks of computing standards-based product costs and variances, management accountants have failed to take an active role in operating the firm.

While the existing literature suggests the management accountant's role is to support the planning, coordination, and control functions in the organization, the primary focus is on control, or bean counting. Hence, rather than being a death knell for the field, AMTs may provide the management accountant with the freedom to fulfill other roles in the organization,

---

[13]R. Howell, "Changing Measurements in the Factory of the Future," *Cost Accounting in the 1990s, Proceedings, The National Association of Accountants,* 1986, pp. 105, 113.

to go beyond supporting management's information needs to becoming part of the strategic decision-making team.

The current problems faced by the management accounting profession are not due to a decreased need for this type of expertise. Rather, management is demanding that its measurement professionals provide more information, on strategic dimensions as well as past performance, in order to achieve manufacturing excellence and a competitive edge in the world marketplace.

The management accountant is being asked to change. The turbulent manufacturing environment dictates a revamping of MAS measurements, an enlarged role for the measurement specialist, and a flexibility in the MAS to match the characteristics of the manufacturing process and the overall organization. No longer solely concerned with tracking inventory costs, the management accountant must begin to measure and report the total set of performance criteria, an all-encompassing role as suggested by Figure 4–3. Instead of a single-

FIGURE 4–3
Emerging Management Accounting Role

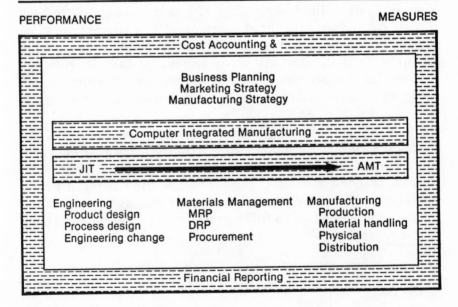

minded concern with product costs, the enhanced MAS will record four main areas: resource acquisition, resource utilization, processing performance, and output measures. An era of challenge and growth awaits the management accounting profession as it assumes a proactive role in the operation of the firm.

> The management accountant can learn to think "Just-in-Time" accounting to keep in step with desirable changes in management style. It is not possible to re-do a whole accounting system in one quantum leap. The management accountant must adjust, however, to changing technology and new management philosophies and be willing to meet the changing demands of information. Nothing is more discouraging than to hear financial people say, "We can't accommodate that change because of our accounting system." Accountants would do well to become familiar with the latest management philosophies and advances in technology so that they can help their firms stay in step and just-in-time as well.[14]

---

[14]R. Seglund and S. Ibarreche, "New Challenges for the Management Accountant," *Management Accounting,* August 1984, pp. 38–46.

CHAPTER 5

THE PROACTIVE ACCOUNTANT

The proactive accountant—to many a contradiction in terms—
is the theme of this chapter. The following quotes graphically
illustrate the common view of the accountant:[1]

> What sort of image does (an accountant) have? Becker, in a light
> mood, described them as most likely to straighten a picture in a
> house as a visitor.
>
> A noted psychologist further reduced the accountants' image
> by labeling them as obsessional, interested in small details, tra-
> dition bound, and noncreative.
>
> A typical auditor . . . (is) a man past middle life, tall, spare,
> wrinkled, intelligent, cold, passive, noncommital, with eyes like
> a codfish, polite in contact but at the same time unresponsive,
> cool, calm, and as damnably composed as a concrete post or
> plaster-of-paris cat; a human petrification with a heart of feld-
> spar, and without the charm of the friendly germ; minus bowels,
> passion, or a sense of humor. Happily they never reproduce, and
> all of them finally go to Hell.
>
> CPA's lament, but have not abandoned, the idea that the
> green eyeshade is still worn by professional accountants.

Accounting, especially management accounting, is a profession
under fire. Some writers suggest that accounting has lost its
relevance. Others perceive it as one of the major causes of the

---

[1]The following quotes originally appeared in E. E. Lawler, III, and J. G. Rhode,
Information and Control in Organizations (Pacific Palisades, Calif.: Goodyear Publishing Company, Inc., 1976), p. 125.

current crisis in U.S. manufacturing. The bottom line is that the value of the management accountant and management accounting itself are being questioned.

Are Accountants Rigid?

Research suggests the stereotype of the accountant is unwarranted. In fact, DeCoster and Rhode found that CPA firm employees scored quite high on sociability, self-control, and flexibility scales.[2] But the concept of a rigid accountant continues.

One explanation for this lies in the basic theme of auditor independence, which is drummed into accounting students. Independence is a critical feature of the audit environment. Who trusts an individual with vested interests to give an unbiased report? The traditional career path of an accountant begins in "Big Eight" auditing, where the socialization process reinforces independence and professionalism.

If an accountant is obsessed, it is with objectivity and fair representation. Dedicated to impartial analysis and a focus on the facts, the accountant serves a vital function in the general economy. Yet, this watchdog mentality does not mesh well with organizational needs. The practicing accountant is in a double bind: professional beliefs are at odds with support of the organization.

The challenge facing these individuals is to make the transition from accountant to manager. Rigidity, the outcome of a dedication to independence and objectivity at all costs, can be conquered. Education and, more important, recognition of the problem provide the first steps in this process.

Holding Individuals Accountable

A basic theme in organizations is the importance of monitoring and rewarding individual performance. In management accounting, the key phrase is *responsibility accounting*. It reflects

---

[2]Ibid.

an overriding belief that individuals should not be held responsible for events or outcomes they cannot completely control. It is an insidious concept, promoting self-interested behavior.

Parochialism describes this single-minded dedication to maximizing departmental, or individual, performance and rewards to the detriment of organizational objectives. Because each individual is rewarded on his or her output, and not that of the overall group or organization, cooperation is shortchanged. The accounting system, which is dedicated to supporting this individualism, may be partly responsible for this undesirable behavior.

Holding individuals responsible for outcomes triggers defensiveness and a search for someone to blame when things go wrong. Rather than asking, "What went wrong?" the question becomes, "Who is responsible?" Accounting does not cause this problem, but it does support its continuation. By becoming proactive managers, accountants can address these issues.

Sticking to the Knitting

Another basic problem accountants face is that they have abandoned, or forgotten, the basic techniques and concepts of their discipline.

Management accounting is the poor cousin to financial accounting. Nowhere is this more apparent than in the courses offered and required at most universities. Focused on supplying auditors for the large accounting firms, most undergraduate programs require little management-oriented coursework. Out of a 30-hour major, 3 to 6 hours of managerial accounting is the norm. Lack of exposure, then, would appear to be one problem facing accountants.

In one or two courses, very little material can be covered in depth. Students are drilled adequately only in standards and variances and come to equate these with the job of the management accountant.

Gradually, this pattern resulted in the blurring of basic concepts. The most significant loss is in the area of traceability. Variable costs have become equated with the term *direct*, while fixed costs have been labeled *indirect* and placed in the overhead pool. If direct labor and materials represent 90 percent of total cost, this is not a big problem. But when cost patterns change, this "laziness" can result in overhead creep.

The Reactive Accountant

Combining professional concerns with educational shortfalls, the rigidity, irrelevance, and defensiveness of management accounting emerge. The role of education and of the basic philosophy of the auditor are barriers to accounting change. Independence leads to isolation for the management accountant. Taken together, these forces create the vision of a reactive accountant who can be more hindrance than help in solving organizational problems.

JIT requires flexible, simple management geared toward a holistic view of the manufacturing process. It is at odds with the reactive accountant's world. What changes are occurring, if any, in leading-edge companies?

STUCK IN NEUTRAL

To change or not to change is the question. Accountants who fail to change can be compared to a car stuck in neutral, going nowhere fast. This metaphor is an accurate description of TeleComm Corporation.

TeleComm was founded and grew to immense proportions in a regulated environment—communications. Its culture was based on the philosophy "furnish the best possible service at the lowest possible cost, consistent with fair treatment of employees and shareholders." In a regulated environment, this philosophy worked. Quality service and products were TeleComm's hallmark, and full, lifetime employment was ensured. Change was not needed or pursued.

Deregulation of TeleComm's markets created a crisis at the organization. While facilities were downsized, cut, and closed, the cultural health of the organization was all but ignored. The company, retrenched along product lines, is voicing a marketing focus but seems unable to shake its regulated mentality.

In this setting, true changes to any aspect of the organization are difficult to achieve. TeleComm remains anchored in deep water with a leaky boat.

Accounting—No Change

At TeleComm Corporation, the appearance of change characterizes its accounting group. After an extensive, time-consuming, and expensive study designed to modernize the accounting system and bring it in line with the demands of the new environment, the company's accounting system remains relatively unchanged. One line item, reflecting actual dollars spent to date, has been added to one overhead report, reflecting the realization that recovery of costs is no longer taken for granted.

As with most existing accounting systems, TeleComm's MAS motivates managers to produce to absorb overhead. One individual noted he could probably cut his direct labor in half on some products, but it would result in an increased burden rate. Additionally, the factory is currently evaluated on output/person, regardless of whether the output is inventoried or sold. End-of-period production swells the joint effect of these measurements.

Don't Make Waves

TeleComm's attempts to innovate continue to meet with premature death, and the new culture arising from the ashes is built on a credo that says, "Appear to change, but in reality stay the same—do not cause waves." This belief is reflected in TeleComm's accountings. The company appears to be stuck in neutral and may be on an uphill grade.

Although the accounting system cannot be blamed for this problem, it is not making any progress toward isolating the problem or reporting on its costs. Accounting works in relative isolation, continuing to spew out complex, lengthy reports of standard costs and variances on production. Managers ignore the reports, yet accounting continues to issue them. It is a vicious cycle with no relief in sight.

LAG THEN LEAD

MicroChip, Inc., reflects a lag/lead phenomenon with respect to accounting change. At times reactive, at others proactive, the accounting process continues to evolve. Although the company has adopted advanced manufacturing technologies, it is just beginning to modify its measurement/accounting systems. These MAS changes, though, go far beyond the immediate information needs of the technology users. Management hopes to use the revised accounting systems to shape behavior rather than merely capture it.

A Pilot Approach

While the JIT process at MicroChip, Inc., is fairly mature, its accounting for this technology is just beginning to emerge. This type of pattern is normal at MicroChip, where one manager said support systems continually lag process improvements. Slow adoption of accounting innovation is not unusual at this company.

The following goals were adopted by MicroChip as part of its focused-factory accounting:

- Determine cost per hour to operate a factory cell.
- Compute velocity of units moving through each cell; the faster units move, the more units can be processed for a given fixed operating cost.
- Produce exactly the right product, on time, with exactly the right mix.
- Eliminate staging inventories and minimize WIP inventory.
- Reduce scrap and other waste costs.

These objectives are part of a move toward a total pull system of manufacturing, with no staging from raw materials through finished goods. This goal reflects the recognition that move and queue time represent increased costs rather than value-added functions.

Focus on Value-Added

This new accounting system focuses on the productivity, efficiency, and effectiveness of each cell. To track performance of each cell, the system measures units scrapped, units completed, units reworked, units for engineering, and velocity of units through the cell, net of process time. Also, total hours worked less non value-added time (e.g., set-up time, downtime, and queue time) are monitored. These measures emphasize cycle time. Hence, the pilot accounting system is embedding value-added concepts into the measurement process.

Figure 5–1 summarizes the critical success factors for MicroChip, Inc., and the manner in which performance at the site will be compared against theoretical optimums. The first part of the figure details seven key measurement areas: unit cost, cycle time, on-time delivery, quality, linearity, inventory, and scrap. These measures are the primary cost drivers and performance factors needed to achieve manufacturing excellence at MicroChip.

A Unique Bonus System

Based on these measures, a bonus system will reward improvement in actual performance against theoretical optimums. This bonus system will be two-sided. When performance on any one of the key measures deteriorates, a penalty will be assigned to the team bonus pool. These penalties offset improvement bonuses, so only net positive changes will result in paid bonuses. This approach will help keep the focus on maximizing total performance rather than manipulating performance and reporting to maximize bonuses. The impact of uncontrollable slippages is one concern, but as the system is implemented these types of problems will be handled on a case-by-case basis.

MicroChip, Inc., in changing its performance measurement system to match the unique characteristics of its JIT production processes, creates a reactive-proactive role for accounting. Many of the changes planned for the accounting

FIGURE 5–1
Critical Success Factors at Microchip, Inc.

SEMICUSTOM ASSEMBLY AND TEST TEAM

Goals:

- Recognize manufacturing excellence and compare to achieved performance (a start point).
- Establish bonus opportunity.

Where:

	Achieved Performance	Theoretical Perfection	Bonus earned % eligible payroll		
			Below Achieved	Achieved Performance	Above Achieved
1. Unit cost	$1.50	$1.00	Negative	0.0	Positive
2. Cycle time (velocity)	7 days	3 days			
3. On-time delivery (demand vs. actual)	60%	100%			
4. Quality (current method) housekeeping	200	0 defect			
5. Linearity (constant flow)	65%	0 deviations			
6. Inventory (segment C.O.S. turns)	15 turns	75 turns			
7. Scrap (yield loss)	30%	0%			

Notes:

- Theoretical perfection could change as new equipment replaces old, cell configuration changes, labor costs go up, etc.
- Theoretical maximum = Maximum bonus when achieved and maintained
- Maximum negative will not exceed maximum positive for any item.
- Achieved performance for start point would be tightened as results improve (suggest bi-annual review). Theoretical maximum would change.

Definitions:

1. Unit cost: Cell cost per hour ÷ theoretical units per hour (unit cost would change as new equipment replaces old, etc.).
2. Cycle time: Theoretical time through cell with no downtime, or minimal.
3. On-time delivery: 100 percent on time and exact mix of demand.
4. Quality: Zero defects
5. Linearity: Measured daily with zero deviation to absolute linearity.
6. Inventory turns: Follows cycle time except for allowed inventory in WIP.
7. Scrap: 0 percent of scrap would require 100% yields.

system may provide the impetus to completely adapt the production process to the JIT concept. The lag/lead role played by the accounting system at this firm provides a unique example of organizational change.

INFORMALLY PROACTIVE

Diesel Systems, Inc., is an interesting example of the mismatch between corporate and division objectives and philosophies. Nowhere is this more evident than in its management accounting system. Diesel Systems is dedicated to the JIT philosophy. To support this dedication, the controller is revamping the management accounting system to match the operating characteristics of JIT manufacturing. But corporate accounting continues to insist that Diesel Systems provide full standard cost-based reports, including variance analysis, on a monthly basis. To meet these conflicting demands, the Diesel Systems controller is maintaining two separate management reporting systems.

The Maverick Controller

The controller describes himself as a maverick. When asked what his role is in supporting management efforts to continuously improve, the answer was, "My job is to serve management, to give them the numbers they need." He is seen as a critical player on the management team, one who is less an accountant than a manager who understands financial (e.g., economic) information.

The controller focuses on comparing performance to plan:

> We do a lot of screwy things to the management control system because of financial accounting requirements. Our key concerns should be: (1) What is the plan? (2) How did we vary from it? and (3) Why? The management control system, therefore, should be providing the roadmap for where you want to be, not where you have been. We have a very elaborate system to justify/track a flaky plan.

Service, not control. Proactive, not judgmental. The controller at Diesel Systems embodies the changing roles and attitudes of leading-edge accountants.

Informal Measures

The constraints placed on the reporting structures at Diesel Systems force the plant-based system into an informal, or renegade, position. Management's view of this dual system is:

> If we can't integrate with the financial accounting system, we'll set up two information systems, because we have to know what it truly costs to run the business.

The objectives and beliefs voiced by Diesel System's management appear rational and appropriate for the JIT manufacturing environment. Yet, these measures are informally tracked and measured, contrary to corporate mandates regarding the "proper" system to use. It is a costly solution for the corporation as a whole, but the company's management firmly believes the new measures and philosophies adopted to date are partly responsible for success at home and abroad.

SIMPLIFY, SIMPLIFY, SIMPLIFY

CompSci Industries is an innovator in accounting change. Believing that it functions in a repetitive manufacturing setting that differs across key dimensions from traditional job-order or process systems, CompSci Industries is tailoring its cost management systems to match the characteristics of its own brand of JIT manufacturing.

Direct Labor as Overhead

Direct labor in a traditional management accounting system is the basis for the allocation process, for efficiency evaluation, and for the product costing model. The concept of labor as overhead, then, appears out of sync—perhaps in error.

At CompSci Industries, direct labor is folded into the overhead pool. Representing less than 3 percent of final product cost in its JIT facilities, direct labor is not important enough to spend time, or money, tracking on the plant floor. On a simple materiality basis, any accountant should reach this conclusion. The idea, though, flies in the face of the traditional accounting model, underscoring the impact of the new manufacturing technologies on the management accounting system.

Matching Accounting to the Cost Flow

The cost flow design of production at CompSci Industries is depicted in Figure 5–2. Elimination of storeroom transactions through direct delivery of incoming materials to the production floor, just in time, is providing the impetus for accounting simplification. The key changes made to date are:

- Development of an accounts receivable matching system that performs automatic matching of the purchase order number to the receipt voucher and invoice.
- "Backflushing" of costs and materials.
- Elimination of work-in-process inventory accounts.
- Creation of a new raw-in-process inventory.
- Elimination of work orders.

Each of these changes reflects a focus on simplifying the accounting system, eliminating transactions not information.

The concept of backflushing of costs and materials, as practiced by CompSci Industries evolved from the elimination of work orders in the proxied continuous-flow setting characterizing JIT manufacturing. While work orders are no longer needed to track product flows through the plant, a mechanism is still needed to trigger the movement of materials into finished good from RIP. Backflushing serves this function, allocating costs incurred to scrap and material variance accounts, as well as to finished goods. Actual material usage variances are generated from a manual tracking of scrap in an on-floor bin.

FIGURE 5–2
Cost-Flow Design

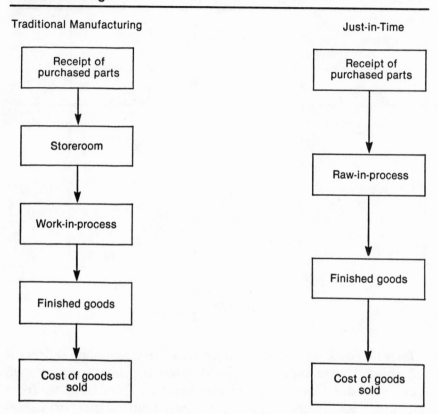

Traditional Manufacturing   Just-in-Time

Replacing Bean Counting with "What If?"

The total impact of these changes to the MAS at CompSci Industries is shifting the accountant's tasks from bean counting to decision analysis and support. Figure 5–3 captures the essence of this process of shifting roles and tasks.

Although some traditional bean counting remains, as financial accounting and some basic management reports are maintained, the emphasis on these types of tasks is drastically reduced. The simplified accountings being adopted make possible simulation and analysis of product profitability, including

FIGURE 5–3
Accounting Department Resource Utilization

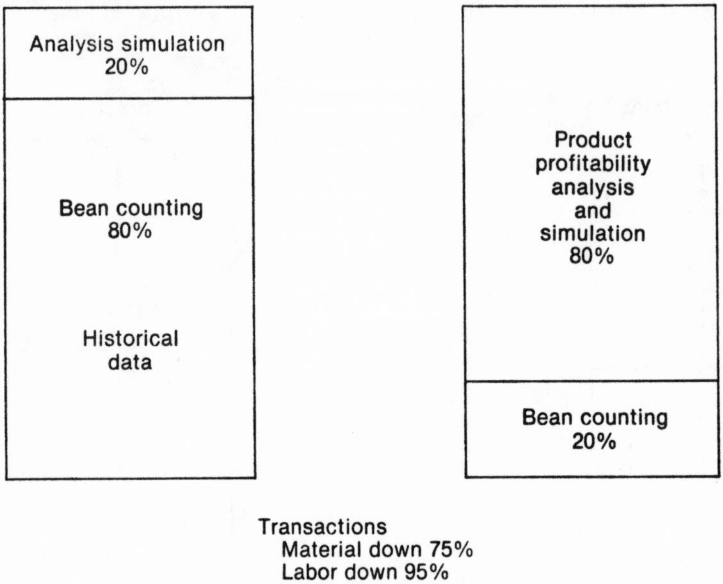

Transactions
Material down 75%
Labor down 95%

life-cycle costs. In the revised systems, management reports a 75 percent reduction in material-based transactions and a 95 percent reduction of labor-related activity. Time freed up from tracking details translates to improved utilization, or value-added functioning, of the accounting group.

Spearheaded by a few highly trained MBAs and a professional accounting staff, the accounting function at CompSci Industries is focusing on "what if," or sensitivity analysis,[3] to support management decision making. Additionally, it is developing simple, yet sophisticated, performance measures matched to the information needs of different functional and

---

[3]Sensitivity analysis is the examination of a problem (and its proposed solution) that takes changes in the key relationships into account. Prior to making a decision, a manager can test how such changes affect the desirability of one solution over another.

managerial levels in the firm. Intelligent use of existing resources and a focus on continuous improvement characterize CompSci Industries' accounting and manufacturing efforts.

A SERIES OF NEW ROLES

General Business, Inc., is adopting both JIT and CIM technologies at many of its major facilities. As with MicroChip, Inc., changes to the accounting system at General Business are being planned, but to date, process improvements exceed accounting change. In fact, the current accounting system reflects a historical costing philosophy using weighted average cost in inventory valuation and both job-order and process costing approaches based on production characteristics. As in most companies, General Business's cost elements are direct labor, direct material, burden (overhead), and engineering. Overhead costs are actually applied to product by computing either labor hours or material dollar value.

In General Business, Inc., the accounting system continues to be ledger driven, but it does interface with material logistics and shop floor control subsystems. Cost is traced using the serial number of products when possible. But because the company still operates in a full cost recovery mode, final costs are fully allocated. Finally, reconciliations to the system are implemented through both the ledger and the logistics/floor control systems, and monthly updates of inventory, product cost, and performance measurement systems are utilized.

Pressures on Existing Costing Methods

General Business's current costing systems are designed for the low-volume, high-cost products of its past. Reflecting a highly integrated manufacturing process, General Business supports a very complex, specialized approach to management. Several key business changes have made the traditional system obsolete:

- Movement to high-volume, low-cost products.
- Increased use of automation.
- Low labor content.
- Dramatic shortening of product life cycles, reflected in shorter release cycles and reduced manufacturing process time.
- High inventory turnover.
- Movement to production management centers encompassing multiple production facilities.

A New Set of Criteria

General Business's desire to ensure its low-cost producer status is reflected in a new set of costing objectives. First, the company is trying to control and reduce its total costs. Such cost-reduction programs are a common tool for cost control. Maturing electronics markets and a drive to secure market share underscore this movement to contain costs.

Second, the cost system is being revamped to match cost and revenues by product, rather than by period or business unit. *Direct charging* is the new buzzword used to describe this type of product-cost model. The implications of this change are significant, tracing to product changes the structure of the cost accumulation process, the types of cost pools maintained, and the definition of direct and indirect costs. The changes provide General Business with the means to compete rigorously on cost and value-added dimensions.

Finally, the company continues to believe that valuing inventory is an important function for its management accounting system. In a high-volume, low-cost setting, these objectives translate into several specific goals for the cost management system:

- Focus on cost drivers and resource requirements.
- Simplify elements of cost.
- Enhance the compatibility between accounting and planning.
- Achieve full cost recovery.

New Roles for Value-Added Accounting

As detailed by a company spokesperson, several new roles are evolving for the management accountant at General Business. First, the accountant is being asked to keep management apprised of competitive pressures. This environmental scanning role is often cited in research literature as important for firms seeking to win the competitive battles of the global marketplace. It is a new role for the management accountant and reflects a drive toward a proactive *management* accountant, an information specialist.

General Business management is asking its management accountant to take a more active role in identifying cost drivers, spearheading cost-reduction programs, and developing expense control techniques. Accounting is being asked to provide a value-added function at General Business. And its accounting group is rising to the occasion.

Figure 5–4 captures the essence of the management accountant's new role to provide more advice and counsel on strategic and business issues. Serving as the information center for the company, the proactive management accountant collects and analyzes data, and provides information (e.g., data endowed with relevance and meaning) to support both operating and strategic decision making.

THE PROACTIVE ACCOUNTANT—A TEAM PLAYER

The proactive accountant is emerging in leading-edge firms across the country. Boldly embracing new roles, with a focus on the "management" part of the title, this new breed of organizational accountant provides the wave of the future. Best depicted as the information specialist on the management team, the management accountant serves an essential function in the organization.

The emergence of a proactive accountant does not mean other individuals on the team are less responsible for pushing

FIGURE 5–4
Management Accounting: An Integrated Approach

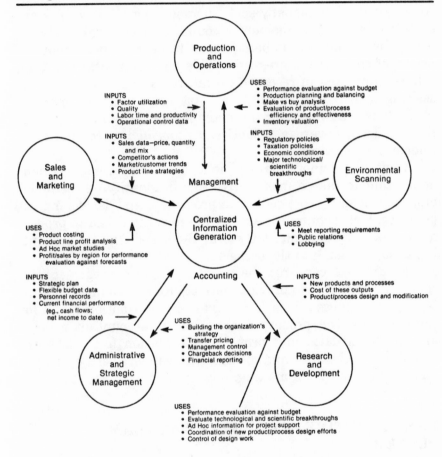

for needed changes in manufacturing, strategic analysis, or accounting techniques. Just as JIT relies on a team mentality for successful implementation, management itself must abandon functional lines and reorganize to successfully meet the challenges of becoming world class manufacturers.

Proactive change entails planning and recognizing the costs and benefits of the alternative manufacturing processes. It is a commonsense approach to management, focusing on automation of only those processes and procedures that are under control and will provide progress on the four criteria of excellence: people, quality, cost, and delivery. A recent comment by William Rotch, professor of accounting and control at the University of Virginia Graduate School of Business, pinpoints the team nature of proactive management.

> The MAS has never been a crystal ball. The farsighted manager—the person—is where the crystal ball resides. Should the controller (e.g., the chief management accountant) be that person with the crystal ball? Sometimes yes, but unless the other managers (production, marketing, research and development) are also crystal-ball gazers, we are asking a lot of the controller to lead the charge.

Perhaps the accountant should not lead the charge, but he or she cannot continue to pretend the battle does not exist. In donning a new hat, with "MANAGEMENT accountant" emblazoned on it, the accountant needs to assume a frontline command position in the battle to regain competitiveness.

CHAPTER 6

NEW MEASURES FOR NEW DEMANDS

The new manufacturing technologies bring new demands for accounting and control measures. In a JIT setting, these new measures are simple representations of material velocity and actual costs. Operational controls, such as statistical process control, visually monitor the flow of components through production. The types of accounting measures evolving mirror the simplicity and responsiveness of the JIT process, supporting the move toward continuous improvement.

REPLACING STANDARDS WITH ACTUAL COSTS

Standard cost models emerged in the 1920s as cost-efficient methods for valuing inventory and monitoring efficiency of the production process. Strong operational controls ensured integrity in the day-to-day management of the shop floor, while the standards-based reports informed top management of general performance and problem spots.

The standards that serve as the basis for this type of system are developed through time and motion studies, and related engineering techniques. A rich body of knowledge has evolved in this area, dealing with how tight or how loose various standards should be to encourage optimal performance, how to analyze actual outcomes against preset benchmarks, and how to generate standards that incorporate normal levels of scrap, rework, and downtime.

Hidden Consequences In Using Standard Cost

As can be imagined, any system that develops numerical evaluations of an individual's performance can lead to manipulation and gamesmanship. Standard cost systems are no exception. Research results suggest that tight but attainable standards are best for motivation, but the actual numbers that fit this definition are anyone's guess.

Here is an example of problems that can occur in the standard-setting process:

> When the time-study man came around, I set the speed at 180. I knew damn well he would ask me to push it up, so I started low enough. He finally pushed me up to 445, and I ran the job later at 610. If I'd started out at 445, they'd have timed it at 610. Then I got him on the reaming, too. I ran the reamer for him at 130 speed and .025 feed. He asked me if I couldn't run the reamer any faster than that, and I told him I had to run the reamer slow to keep the hole size. I showed him two pieces with oversize holes that the day man ran. I picked them out for the occasion! But later on I ran the reamer at 610 speed and .018 feed, same as the drill. So I didn't have to change gears—and then there was a burring operation on the job too. For the time-study man I burred each piece after I drilled and reamed, and I ran the burring tool by automatic feed. But afterwards, I let the burring go until I drilled 25 pieces or so; and I just touched them up a little by holding them under the burring tool.[1]

Invalid data reporting is just one of the undesirable outcomes of the standard cost system. Negative attitudes, which emanate from the above quotation, are also a problem. As tools of control, standard cost systems often lead to such dysfunctional consequences.

---

[1] E. E. Lawler, III, and J. G. Rhode, *Information and Control in Organizations* (Pacific Palisades, Calif.: Goodyear Publishing Company, Inc., 1976), p. 89. This quote is a reprint from W. F. Whyte, *Money and Motivation: An Analysis of Incentives in Industry* (New York: Harper, 1955), p. 18.

Enforced Mediocrity

There are obvious discrepancies between the standard and actual performance. The above quotation highlights this phenomenon. But the actual problem is much deeper than faulty reporting and gamesmanship. Standard cost systems are designed to detect deviations from the preset standard. Since these standards include accepted levels of slack, waste, and downtime, they represent a form of enforced mediocrity. Improvements in processes are not encouraged, because improvements yield variances and trigger the management by exception system to focus on the area. Production personnel do not traditionally seek ways to attract management attention. Their goal in this setting is to minimize deviations and avoid variances.

Enforced mediocrity has no place in the continuous improvement setting characterizing JIT manufacturing. Whereas standard cost-based reporting systems promote a "minimize deviations from standards" mentality, JIT procedures work to minimize variance in the process, not in the measurement system. This is a subtle point. "Meet standard" is in direct contradiction to "continuously improve." "Minimize variances from standard" is significantly different from "remove all forms of variance from the process." Standards that contain waste, in whatever form, mask the sources of variation. In pursuing linear production, JIT balances flow, isolates and eliminates waste, and provides the basis for effective manufacturing. To match this change in philosophy, the management accounting system has to undertake major change.

Standards as a Basis for Costing Products

If the standard cost system, as currently used in most companies, provided accurate product cost data, the dysfunctional behavior suggested above would not be as deadly. Yet, Chapter 4 cites the inaccuracy of product costs as one of the major complaints leveled against the existing management accounting system.

We cannot say for certain that the basic structure of the standard cost system is responsible for faulty costing of products and processes. As argued earlier, in many cases, the inappropriate use of existing systems and techniques appears to be at the heart of the problem. No matter what the basic cause, standards fail to determine the actual cost of producing a product. The following comment by Diesel System's controller sums up this problem:

> What we are actually trying to accomplish is to develop a system which captures the actual unit cost of producing an item—it would include everything, quantitative and qualitative, necessary to produce it. In the past, we could tell performance against standards up to a mil but we couldn't tell the actual cost of anything. I feel this is a bad basis from which to run the business. Our approach is that we're trying to find out where we are.

Misreporting, negative attitudes, and faulty costs—all undesirable characteristics for the basic information system of the organization. What changes are being made to address this problem?

UNIVERSAL ADOPTION OF ACTUALS

Companies are replacing engineered standards with actual costs at all the sites contacted during this study. Frustration with a standard cost system in which product costs can vary based on volume of the total plant, rather than on characteristics of the product itself, triggered Diesel Systems management to move to actual costs. At CompSci Industries, actual quantities used and costs incurred are utilized to evaluate the production process. While a standard cost is maintained, careful inspection reveals it is a rolling average of actual costs rather than a true engineered standard.

The information-processing capabilities provided by the computer make tracking actual costs to production simple and cost effective. Using actual costs means the efficiency and effectiveness of the process can be monitored against prior period performance and theoretical optimum. In tracking changes in actuals, the message to continuously improve is captured and transmitted through the management accounting system.

There Are Dangers

In designing its incentive system, MicroChip, Inc., plans to provide bonuses for process improvements. The plan is constructed to prevent improvements on one facet of the process at the expense of other areas. This two-sided bonus system assigns penalties to the team bonus pool when performance on any key measure deteriorates. These penalties offset improvement bonuses, so only net positive changes result in paid bonuses. This approach focuses on maximizing total performance, and so discourages manipulation of performance and reporting to maximize bonuses.

The danger in the new system is that the target—tight-but-attainable standards—is not embedded in this approach. At some stage, further improvements on prior performance become difficult. If the production process being monitored is static, both in terms of product and technology, improvements will eventually stop. Potentially, this increasing pressure for performance may trigger unforeseen responses from employees. Given the newness of these various approaches, any projection of final impact is simply a guess.

Rolling averages of actual costs are a far different form of benchmark than a predetermined standard with built-in waste. Reflecting history, rather than the gamesmanship skills of the machine operators, rolling averages both promote and reward efforts to reduce waste and other non value-added aspects of the production process. But since most of the distortion in product costs arises from the overhead allocation process, these changes provide only part of the answer. Chapter 7 examines the overhead and inventory issues arising in a JIT setting.

TIME: A CRITICAL FACTOR IN VALUE-ADDED PRODUCTION

The goal of JIT manufacturing is to maximize the movement of materials through the production process. This goal is also embedded in the concept of maximizing value added in the production process. The total time spent in manufacturing can be broken into several key components:

$$\underset{\text{time}}{\overset{\text{Total}}{\text{cycle}}} = \underset{\text{time}}{\text{Process}} + \underset{\text{time}}{\text{Move}} + \underset{\text{time}}{\text{Wait}} + \underset{\text{time}}{\text{Inspection}} + \underset{\text{time}}{\text{Set-up}}$$

Of these components, only process time is truly value-adding.

Focusing on minimizing move, wait, and inspection time is the critical mindset change associated with JIT manufacturing. Seeing the production process in a holistic light exposes the buffers, and waste, currently buried in the manufacturing processes of most major companies. An accounting system that measures only the efficiency of individual operations does not provide the means to identify the magnitude, or cost, of these buffers.

JIT manufacturing redefines the primary unit of analysis from individual activities to the entire production process. Move and queue time, unimportant in determining labor-based efficiency measures, are critical in a drive to maximize the velocity of materials through a plant. In this setting, total cycle time, not individual efficiency rates, dominates.

Given time's primary role in JIT manufacturing, time-based measures are as important as those based on cost (e.g., standard or actual). Several of the companies visited during this project are factoring time into their performance measurement systems.

MicroChip, Inc.: A Focus on Value-Added

MicroChip, Inc., is moving to a total pull system in its JIT manufacturing, with no staging of components (WIP) from die cutting to finished goods. This goal reflects the company's recognition that move and queue time represent increased costs, rather than value-added functions. Two key measures of time are being used: (1) velocity of units through the JIT cell, net of process time, and (2) total hours worked less non value-added time (e.g., set-up time, downtime, and queue time). These activity measures emphasize minimizing total cycle time. They also embed value-added concepts into the measurement process.

Value-added cost collection at each of the JIT cells at MicroChip is computed based on a theoretical estimate of optimal process time, as follows:

$$\begin{pmatrix} \text{Cost of} \\ \text{good units} \end{pmatrix} = \begin{pmatrix} \text{Process} \\ \text{theoretical} \\ \text{time} \end{pmatrix} \times \begin{pmatrix} \text{Cell} \\ \text{cost per} \\ \text{hour} \end{pmatrix}$$

This approach tracks non value-added cost variations through the use of actuals.

In tracking the actual variances at the individual cells, MicroChip focuses on two elements of cost: purchase price variance to piece part standards and value-added variations. The first variance eliminates purchasing performance and uncontrollable market price movements from the production evaluation process. It also allows MicroChip to use a pure form of value-added measures for monitoring internal performance because material costs are kept constant across different settings.

The value-added measures are designed to monitor spending variances (e.g., comparison of budget to actual for level of output); lost time, or capacity/volume variances, where lost time is defined as actual lost time multiplied by the standard cell cost per hour; and velocity, or efficiency/speed, variances (e.g., planned versus actual time at actual cell cost/hour). MicroChip is combining value-added time with actual cost to provide the framework for its focused-factory accounting.

Keeping Beat at TeleComm Corporation

The JIT facility at TeleComm Corporation is also using value-added concepts in designing its new accounting systems. The assembly process is broken into nine major operations, and actual assembly time is divided into *beats*—2½-minute intervals used to synchronize the process. Out of the total cycle time of four hours, 34 percent is spent in actual production (value-added), while the rest is spent in queue and move.

TeleComm is monitoring this arrangement through two specific performance measures—velocity and hours per machine. The first measure is compared against the theoretical optimum (as currently conceived) of four hours. The second, measured specifically as total hours of machine time per units produced, provides the basis for evaluating actual value-added time in the cell. As at MicroChip, Inc., the role played by time in causing cost has been recognized and is being addressed through new measures and processes.

MAJOR MEASUREMENT ISSUES

Adoption of actual costs and time-based performance measures address many of the perceived shortcomings of the management accounting system. But these two changes in isolation do not alone capture the entire realm of improvements and changes that characterize a JIT manufacturing process. Several other new areas of measurement are: obsolescence, standardization of parts, and total defects.

Tracking Obsolescence

Two issues emerge in discussing obsolescence, or technology decay, versus traditional depreciation methods. First, as technology changes, the useful capacity of the affected area drops to zero. Yet the depreciation of this equipment is usually incomplete. This leads to a write-off of obsolete equipment, which is both a financial loss to the firm and a barrier to new technology adoption.

Temporary Retirement of Assets
A related issue is the temporary retirement of capital equipment and technologies, such as the robots used in TeleComm Corporation's keyboard assembly area. As reported by TeleComm representatives:

> We need to figure out, through creative accounting, how to put a big plastic bag over a hunk of capital in place. If you ever need

to, you put it back on the books. The current methods increase burden rates to accommodate the costs of temporary over-capacity.

This raises the issue of useful life versus use, or the need for a units-of-production (e.g., an activity-based depreciation method) mentality in charging out technology and capacity costs. The plastic bag suggested here holds productive potential, not forgone production. It is an opportunity cost rather than a sunk-cost approach to capital equipment management.

Life-Cycle Accounting

In a life-cycle approach to accounting, the useful life of technology is defined not on total lifetime output but on planned usage. By attaching these technology and depreciation costs directly to the products benefiting from them, knowing the true costs of products, as well as total cost recovery, is more likely. Whereas none of the firms visited had specific measures designed to combat obsolescence, many firms are beginning to explore this issue in revamping their overhead systems.

The major phases in a product life-cycle model are marketing, technical investigation, prototype, pilot, manufacturing, process control, maturity, and end-of-life.

Beginning with a predetermined need, or unfilled niche, in the market, the company develops a new product. As suggested earlier, more than 90 percent of the final product cost is locked in before manufacturing begins because of product and process design decisions. Yet the meter in most accounting systems does not begin ticking until the manufacturing stage. Prior costs, expensed in line with financial accounting conventions, are omitted from the determination of product costs.

If properly planned, technology in operation at the start of production of the product can be charged out equally to all items produced. Conversely earlier output, usually benefiting from higher profit margins, can be allocated larger portions of these costs. Contrary to traditional depreciation, this allocation plan is not driven by the expected life of the equipment but by the expected units of product to be produced with it.

Refocusing cost collection and distribution from periods to products significantly impacts the accounting methods employed and the types of cost pools adopted. In no other area is the new management accounting system in such direct contradiction to the basic structure of financial reporting.

Financial reporting seeks to evaluate the profit-making activities of an enterprise for a specified period. Life-cycle costing, on the other hand, focuses on accurately identifying all the costs caused by the production, *over time,* of a product or product line. Although not solely a result of the move to JIT manufacturing, this trend toward life-cycle costing reflects the holistic view of manufacturing embedded in JIT and related flexible manufacturing approaches.

First-Pass Part Measures

The renewed focus on quality that accompanies the adoption of JIT manufacturing is witnessed by the formal measurement and reduction of defects and rework. At Diesel Systems, Inc., this characteristic is informally measured and controlled by management through the statistical process controls in place on the plant floor. Good process controls eliminate the need for further tracking.

TeleComm Corporation maintains several quality-based measures in its formal management accounting system:

- First-pass yield for each process versus first-test yield.
- Daily scrap tickets.
- Good units produced versus planned output.
- Initial operability, as tracked by customer service.

First-pass yield is also a key measure at MicroChip, Inc. Measuring the number of units produced right, the first time, reflects the recognition that quality does not cost money—rework and scrap do.

Comparing good units produced to planned output reemphasizes the goal to produce to demand (e.g., production linearity). This linkage emphasizes the pull nature of JIT manufacturing and suggests excess production is not the way to hit the good units metric. This measure is expected to be below one or, at perfection, equal to one. Producing 120 units to get 100 good ones is discouraged because of the risk of overshooting the mark on total planned production.

Goal: Zero Defects

The goal of zero defects is critical in ensuring linear production. Each of the companies visited recognizes this relationship and is in some manner tracking its progress toward this goal. At CompSci Industries, the presence of a defect stops the line. This suggests that within prescribed tolerances, zero defect production is ensured. Whether through operational controls or management accounting-based measures, the benefits of quality in product and process are tracked.

Standardizing Parts and Processes

The standardization of products and components is inherent in the move toward product families as the focal point for organizing the production process (the basis of group technology). Ideally, the company maximizes commonality of parts and production to a stage where customization can be a final step rather than the basic feature of the manufacturing process.

Diesel Systems, Inc.: Major Progress

Diesel Systems, Inc., gears its production process toward producing subassemblies that can be rapidly converted and completed once an order for a particular model of fuel injector is received. Maintaining this inventory of subassemblies enables Diesel Systems to effect product mix changes in two to three days. Currently, 25 models of automotive fuel injectors are produced. Out of 180 different parts for each, roughly 165, or 92 percent, are standardized. In the older markets, not yet converted to JIT, standardization remains at 40 percent of total components, with more than 3,000 models actively produced.

Diesel Systems does not directly measure the benefits of decreased complexity, such as the standardization of components. Instead, it tracks costs caused by a failure to standardize (e.g., the costs of complexity of product design). Examples are lead time to effect product mix changes (responsiveness), inventory turns, and total lead times. Not each new demand for information in a JIT setting can be addressed with direct measures. Instead, companies are using cost drivers to measure how improvements on the drivers affect various aspects of the costs caused by such elements as excessive product complexity.

AMTs Require Standardization
General Business, Inc., is tied to standardization by the very nature of the technologies it is installing. Robots are not as flexible as people unless excessive programming is used. Standardization and its benefits begin in the initial design of the product. Increasing use of CAD/CAE by U.S. manufacturing firms suggests many of these companies are recognizing the benefits achieved by designing simple, standard products. As suggested above, these efficiency and effectiveness improvements can be tracked through their impact on areas of cost.

INTEGRATED MEASURES FOR HOLISTIC MANUFACTURING

While a number of measures can be used to track different performance characteristics in any manufacturing setting, the holistic manufacturing concepts embedded in JIT are leading to the use of integrated measurement systems. The efforts at CompSci Industries present a comprehensive example of this trend.

A Total Performance Measurement System

CompSci Industries is combining the measures suggested here into a total performance measurement system that integrates the measures used to control the entire production system from receipt of the customer order to delivery. At the division president level, only key marketing and financial measures are used to evaluate performance. Marketing results are keyed to company growth versus market growth, company growth versus market plan, and the order fill rate. Total costs of material shipments and overhead, division earnings, asset turnover, and capital requirements are the key financial details used to evaluate performance at this level.

An Emphasis on People

At the general manager level, the four criteria for excellence discussed in Chapter 2 begin to be used. Gross measures of people, quality, delivery, and cost performance are captured and reported. CompSci Industries is people-oriented and emphasizes management by wandering around. Several CompSci Industries brochures underscore the belief that people are key to the firm's success. As noted in one company document:

> Our people are our most important resource and the managers have direct responsibility for their training, their performance, and their general well-being. To do this a manager must get around to find out how their people feel about their jobs and what they feel will make their work more productive and meaningful.

One of the key performance measures at the general manager level is a daily survey of employees.

Measuring the Other Criteria for Excellence

On the other three dimensions of excellence, the measures used reflect the high level of control inherent in the existing production process. Quality, while a trademark of the company's products, is reported only as an annual failure rate because it keeps top management informed of any major changes. Delivery performance is evaluated based on on-time delivery rates. Finally, the general manager is held responsible for three components of cost: manufacturing product cost, fixed cost levels, and related variable costs.

While the detail expands as we descend the pyramid, performance measurements are a direct application of responsibility accounting translated through the company's people-oriented culture. Pursuing a belief that initiative and creativity are fostered by allowing the individual freedom of action in obtaining agreed-upon objectives, measures are tailored to the lored to the individual's goals as well as key company success criteria based on strategic goals.

Looking Toward the Future at CompSci Industries

The total performance measurement system is designed to integrate activities of individuals throughout the company and to support movement toward continuous improvement and attainment of its strategic objectives. But the system is not complete in management's eyes. CompSci Industries is seeking to redesign this system to provide cost and other measures that support resource allocation decisions consistent with its manufacturing environments. The following checklist reflects the system's future requirements:

- Product design.
- Process layout.
- Process routing.
- Make versus buy decisions.
- Vertical integration.
- Process design and control.
- Capacity planning.
- Machinery and equipment planning.
- Land and building planning.
- Strategy development.

These decisions are made in various manufacturing environments. Each of the decision information demands is expected to be met in a simple, flexible, and real time setting, in a manner consistent with the manufacturing environment, whether continuous flow, multiproduct, or service (or supplier) to other subsidiaries or plants in the company.

NEW MEASURES FOR NEW DEMANDS

This chapter detailed key areas where the performance measurement, or management accounting system, is being changed to reflect the characteristics of JIT manufacturing and its focus on the continuous elimination of waste. A brief summary of this discussion is:

- Actuals are replacing standard costs at all sites visited.
- Value-added characteristics, such as total process time, are being tracked more intensely.
- Obsolescence, elimination of defects, standardization of components, and velocity of materials are areas being targeted for measurement and improvement.
- The final goal is a total performance measurement system that integrates organizational activities across various managerial levels and functions.

The management accounting system of the future needs to incorporate many factors currently not measured and to focus on transactions that cannot be directly tied to debits and credits in the general ledger. By recognizing that intangibles, such as quality, cause tangible shifts in overall costs, progress is being made at leading-edge firms.

> Somewhere we need to be able to quantify the intangibles, such as better quality control, decreased inventory, decreased scrap, decreased queuing, and perhaps things like competitive advantage. We know these items are going to pay off eventually, and I really think you could make a good stab at quantifying them.
>
> Controller, Diesel Systems, Inc.

CHAPTER 7

OVERHEAD: ATTACKING COSTS THAT DON'T ADD VALUE

Overhead allocation methods have an overwhelming impact on estimated product cost. As suggested in earlier chapters, both the number of cost pools and the activity measure or measures employed to apply this cost to product affect the accuracy of the final cost estimate. The word *estimate* is used for a reason: "true" product cost is the goal, but one the accounting system cannot achieve without excessive information collection and processing costs. The operationalized goal of product-costing techniques, therefore, is to generate the most accurate costs within the cost constraints imposed on the system.

Apportion versus Allocate

Over the years, the basic language of management accounting has lost its clarity and precision. Nowhere is this more evident than in the area of overhead techniques. Today, all overhead is "allocated." *Apportionment* is a lost term in the accountant's vocabulary. Re-adopting the word and the concepts it embodies is an essential first step on the road to regaining relevance.

The key difference between the concept of allocation and apportionment is the *homogeneity* of the underlying cost pools. In apportioning, the accountant seeks to group all costs *caused* by a common cost element. The focus is on creating homogeneous pools of cost that can then be applied to product, or process, based on the appropriate activity metric, or driver.

Machine Labor-Hour

An example of apportionment is the machine labor-hour approach suggested by A. Hamilton Church in the early 1920s. Defining a machine as the basic pool, all costs associated with running that machine are placed in the appropriate "bucket." Each product moving through that machining center is then charged with its portion of this cost.

In defining the cost pool around a machine, or machine center, many elements enter into the picture, including:

- Direct and indirect labor required to maintain production in the area.
- Depreciation on the machine and associated tooling.
- Supplies used in the production process.
- Power and other utility costs.
- Supervisory time/cost.
- Floor space "rent."

Each of these costs is caused by the machine. And each unit moving across the machine picks up its share of this cost. Estimates are still involved in determining which costs to place in the pool and how much of this cost to apportion to individual units passing through the machine center, but they are reasonable estimates based on fact rather than an arbitrary lumping of cost on product.

Then What Is Allocation?

Obviously, not all costs can be placed into these homogeneous cost pools. The best example of one such set of costs is those of the accounting department itself. Many of the factors that cause cost in this area are external, such as tax compliance and related regulatory effects. Although some of the costs can be more directly traced to production (e.g., time spent in reconciling direct labor costs), it is a tenuous tie.

Pooling each of these truly indirect costs creates a *heterogeneous* bucket of expenses that, according to Financial Accounting Standards Board and Securities and Exchange Commission rulings, must be *allocated* to product in arriving at full costs. These costs are arbitrarily assigned to product to meet the requirements of these regulatory bodies. Most managers know that full costs are an inappropriate basis for most decision making. Given the fact that regulations, and not management information needs, drive the full product-cost models, it is obvious that determining full costs should not be the objective of the management accounting system.

The term *allocation* implies arbitrary. The pool of cost is heterogeneous, so utilizing one causal agent for assigning cost to product results in inaccurate product costs. Allocation, not apportionment, distorts the final product cost because of this arbitrariness in cost assignment. Since they do distort the final cost number, allocations should be minimized.

The more a company utilizes allocation in its management accounting system, the more distorted and inaccurate the resulting product cost figures become. Given that strategic product line and pricing decisions are often based on these costs, extensive allocation can trigger the garbage in–garbage out phenomenon and, in the extreme, destroy the profitability and viability of a company.

How Many Cost Pools?
The above discussion suggests that more cost pools, indicating more homogeneous groupings apportioned on the primary causal agent, are better. When does it become too expensive to maintain another layer of cost pools? How can a company determine when it has enough pools? Two factors enter into answering this question: information collection/processing costs and the percentage change in final product costs.

The second factor, the impact of increasing the number of pools on final product cost numbers, suggests the management accountant employ a form of sensitivity analysis in arriving at the optimal number of homogeneous cost pools. By refining cost tracing to pools, the analysis focuses on identifying the point at which increased precision in the tracing process has little or no impact on the final product cost. At this stage, any further detail represents an unnecessary expense. Determining the actual costs of increased detail can be difficult. Incremental analysis provides some help. Beginning from some estimate of the cost of maintaining the existing system—in terms of labor costs in the accounting group and related support (i.e., computer costs)—enhanced systems are examined to determine their resource implications. The analysis of incremental costs then proceeds much like that used in capital budgeting. The additional information cost of adding refinement to the cost system is evaluated against the improvements made in the output—the estimated product cost.

This approach provides only a rough approximation of the costs and benefits of improving the accuracy of the product cost system. But being approximately right, versus precisely wrong, reflects a value-added approach to the management accounting system and its redesign.

Management Accounting: An Art
In implementing this type of approach, the management accountant must refine and utilize basic analytic tools currently missing, rusted, or broken. Rather than counting beans, the accountant focuses on understanding the basic processes and products well enough to develop accurate, homogeneous pools for apportionment with minimal information tracking, reporting, and cost. The goal becomes creating a value-added accounting system that provides a flexible, relevant, and timely basis for managerial decision making.

The age-old argument of whether accounting is an art or a science is partially settled here. The quest for apparent precision, or rigor, in the accounting process resulted in inaccurate product costs, irrelevant information, and faulty signals to management regarding product costs, resource utilization, and related items. Reflecting goal displacement, or "means-ends" inversion, management accounting placed precision in the debit-credit machine ahead of precision in the output of that process. In attempting to become a precise science, accounting moved into the realm of the irrelevant.

Sensitivity analysis, apportionment, cost-benefit analysis, and incremental analysis are tools that exist in the textbooks. They represent the art of management accounting. It is an art which, when mastered, provides a precise, useful, and relevant landscape of the organization.

Abandon Precision for Relevance

The focus on precision in measurement must be abandoned. It must be replaced with a concern for the relevance and usefulness of the output. Art requires professional dedication, talent, and knowledge to bear fruit. Science seeks technical precision, which in accounting translates to techniques easily programmed into a computer. By returning relevance, and art, to management accounting, the professional status long sought by management accountants may finally be attained.

PROGRESS IN THE FIELD

What types of changes in the overhead area are evolving in practice? Each company visited is using a different approach to address overhead. The techniques and systems being developed reflect the strategic objectives of each firm.

Diesel Systems, Inc.: Refine and Isolate

Diesel Systems, Inc., has replaced its traditional full absorption cost-based reports with a contribution margin reporting format. A typical report generated by the new system is presented in Figure 7-1. The key characteristic of the revised management accounting system used by the plant is the *nonallocation* of overhead items to product.

The income statement used to evaluate the plant's performance (remember that corporate management has not bought into Diesel Systems' new accounting system) matches the direct costs incurred in the production process against associated revenues. The remaining indirect, or overhead, costs are presented as separate lump-sum line items. This contribution margin approach focuses management's attention on items it controls (i.e., overhead such as administrative costs), while simplifying and refining the reporting process in the plant.

Don't Poison Your Product Costs

Management's objective at Diesel Systems is to change the accounting system's role from that of historian to one that aids

FIGURE 7-1
Income Statement (January, 1986 in $000's)

	Current Month					Year-to-Date				
	Total	Pump	Nozzle	Filter	Other	Total	Pump	Nozzle	Filter	Other
Net sales										
OEM	2,000	1,000	500	200	300	2,000	1,000	500	200	300
Service	2,000	1,500	300	100	100	2,000	1,500	300	100	100
Total net sales	4,000	2,500	800	300	400	4,000	2,500	800	300	400
Direct costs										
Material										
OEM	800	400	150	100	150	800	400	150	100	150
Service	700	500	100	50	50	700	500	100	50	50
Variance	10	10	(10)	5	5	10	10	(10)	5	5
Total material	1,510	910	240	155	205	1,510	910	240	155	205
Factory										
OEM	725	400	200	75	50	725	400	200	75	50
Service	525	300	100	75	50	525	300	100	75	50
Variance	30	10	10	5	5	30	10	10	5	5
Total factory	1,280	710	310	155	105	1,280	710	310	155	105
Direct profit	1,210	880	250	(10)	90	1,210	880	250	(10)	90
Direct profit percent	30.3%	35.2%	31.3%	-3.3%	22.5%	30.3%	35.2%	31.3%	-3.3%	22.5%
Factory overhead	300					300				
Selling & admin.	100					100				
Engineering	100					100				
Tappet	50					50				
LIFO Setup	10					10				
Europe P/L	(2)					(2)				
Miscellaneous	252					252				
Income before taxes	400					400				
Return on sales	10.0%					10.0%				

in the planning process—from a focus on where it has been to where it is going. Resources are concentrated on budgeting overhead items, ensuring overhead remains at or below budget, and determining direct and "semidirect" (e.g., apportioned) costs for controlling operations. Control is achieved through careful planning and utilization of process controls.

Standards determine flexible budget targets for different levels of capacity utilization and isolate areas that can be improved. The management philosophy related to business reports is summed up as follows:

> Why poison your outlook with full costs? We focus on controllable factors at the reporting level for managers. Putting overhead in broad buckets facilitates forecasting. So, volume-based accounting measures are still used in forecasting and planning. . . . We look for a pure control approach. Namely, for each level of reporting we focus in on control and traceability. What we'd like to do is look at the business. We know the direct costs, and know the profit percentage we can make on these items; therefore, we know how much we can afford in overhead. Our reporting system focuses on identifying those items we can change. So, at the top level, there's little that can be done to direct costs (although on the plant floor these can be cut), but we do need to keep an eye on overhead items.

Gearing Information to Use

Diesel Systems, Inc., focuses on control. This is emphasized in the next chapter and so will be only briefly discussed here. Given a desire to hold areas accountable only for those elements of cost they can control, Diesel Systems adopted contribution margin accounting as its basic system. Full costs must be utilized to value inventory, but this is not the goal behind accumulating the raw data and processing it into information. The system reflects management's philosophy of responsibility accounting as the primary task performed by the management accounting system.

Diesel Systems is revamping the accounting system to reduce the pool of indirect, untraced overhead in hopes of developing more refined cost pools to apportion more overhead to product. Accurate products costs are critical for Diesel Systems' survival in a highly competitive market. Best described as semidirect costs, these apportioned charges seek to bridge the gap between direct and indirect costs to more accurately reflect the resource consumption of various processes and products.

Two Drivers for Overhead Allocation

CompSci Industries, a leader in the development and application of JIT manufacturing processes and procedures, is less progressive in its attack on overhead. Its major change in this area is development of a second allocation basis—material value. In general, it allocates production overhead on direct labor hours and material and support overhead on material value. Production overhead includes all direct labor (wages and salaries), supplies, expensed equipment, depreciation, and service department costs. Material overhead is defined as purchasing, stores and material handling, engineering changes, production planning, materials quality audit, and traffic costs. Finally, support overhead includes administration, management, and engineering area charges. While having only two overhead bases is a long way from the goals of JIT accounting, or activity accounting, it is an improvement over using direct labor to assign material-based costs.

In folding direct labor costs into a production overhead pool, CompSci Industries created a heterogeneous pool of cost that is *allocated* to product based on direct labor hours. In many respects, then, direct labor is still tracked, although not in a detailed matching of individual labor costs to product.

Material Value as a Driver
The second activity measure added to CompSci Industries' system to reflect the characteristics of JIT manufacturing is material value. Two overhead pools are charged out on this basis. The first, *material overhead,* is an apportionment of material-based costs on the dollar value of that material. Included in this pool are purchasing costs, stores and material handling, engineering change-related costs, production planning, material quality audit, and traffic. Each of these support areas in-

curs cost to move material. This apportionment of material-based costs captures the essence of material velocity.

The second group of costs charged out on material value constitutes an *allocation*. The support overhead pool (e.g., administration, management, and engineering) is allocated to product based on material value. This is the same charging basis, but this time the process is an allocation rather than an apportionment. Therefore, the second use of material value is likely distorting product cost, whereas the first is not.

As Always, Change Brings Dissatisfaction

While CompSci Industries management firmly believes the two-driver system is superior to prior overhead treatment, even the change described here generated dissatisfaction at one site. A conversation with a design engineer revealed the implications of focusing on material value.

The engineering manager described a recent situation where, in an attempt to improve manufacturability, a product was redesigned to utilize one integrated component in place of 50 smaller parts. The material value, or cost, of the replacement unit significantly exceeded the cost of the pieces it was to replace. But labor hours were to be cut, quality enhanced, and overall reliability of the product improved. The qualitative benefits matched CompSci Industries' strategic manufacturing goals. When the material value-based overhead was assigned, though, the bottom-line impact was an increase in the final product cost. The change was not made.

The design engineer's frustration was easy to see. In pursuing design for manufacturability objectives, it was appropriate to simplify the product in question. But because of changes in the overhead charging structure, the product was left unchanged. This type of mixed message can severely impact the morale and dedication of employees.

Moving Beyond Material Value

CompSci Industries is not stopping its efforts to adopt activity-based accounting with the current two-driver system. Corporate attention is focused on developing a multidriver system that attempts to zero in on "true" product cost. As its markets mature and cost competition begins to erode market share and profits, CompSci Industries must understand which factors cause product costs to change in order to optimize manufacturing.

Manufacturing Cells as Cost Pools

While CompSci Industries is pooling costs along traditional material, labor, and administrative overhead lines, MicroChip, Inc., is redefining its primary cost pools to reflect its manufacturing cell configuration. In tracing costs to cells, the management accounting system focuses on improving the effectiveness and efficiency of team output.

Two main elements in the development of a costing system are: the basic costs that are pooled, and the metric or activity measure used to charge those costs to product. At CompSci Industries, the cost collection elements by cell are:

- Team labor, including production, maintenance, and quality assurance individuals.
- Die costs and purchased parts, charged at standard cost through the bill of materials.
- Equipment depreciation within the cell.
- Maintenance and quality assurance support costs (e.g., those not directly traced by charge).
- Power usage.
- Space cost.
- Other operating expenses.
- Administration, engineering, and related support charges.

These cost elements are pooled and then divided by the total number of hours the cell is operated to derive cell cost per hour of operation. This number is computed monthly and tracked for evaluation of continuous improvement goals.

Similar Activities—Similar Costs

The *place* where costs are incurred defines homogeneity of costs in a JIT setting. The costs entering the pool appear heterogeneous, but all are traceable to a specific cell. The charging basis, cell cost per hour of operation, represents an apportionment based on where the costs are incurred.

This system matches the cost flow assumptions of the accounting system to the physical characteristics of the process. The manufacturing cells in JIT manufacturing are islands of pure value-added processing. The value added is more than direct labor. It is every cost incurred to support the cell.

Recognizing the Interdependence of Cost Elements

By tracking the patterns in the cell cost per hour of operation, MicroChip, Inc., can eliminate or reduce nonessential elements of this package of costs. Key to this approach is the interdependence of the total components of cell cost. In most management accounting systems, each cost is tracked independently. Interdependence is not analyzed or reported.

In moving to the cell-based cost system, MicroChip, Inc., can analyze the impact of change on the entire system. If maintenance is reduced, what happens to total operating expenses in the cell? If reduced maintenance results in increased downtime, the cell cost *per hour of operation* may increase. Even without understanding precisely how these various cost elements are linked, management can determine if changes improve the overall performance of the cell.

Inventory Valuation—A Cell Approach

One goal of the new accounting system being tested at MicroChip is to simplify inventory valuation procedures by modifying them to utilize cell costs. A company spokesperson described the preferred way to achieve this goal as a three-step process:

1. Count the units in each cell at the end of the last shift in the month.
2. WIP = units × actual cell costs/hour.
3. When cycle time is fast enough, expense all factory costs each month.

The inventory tracking unit being used to simplify accounting will initially serve as the source of unit costs. Because of the interim cell configuration in the plant, several functions will still be charged out at standard. The final WIP valuation will be a hybrid of standards and actuals.

Two-Stage Overhead Allocation—The Goal

The final goal of the new accounting system at MicroChip is to adopt a two-stage overhead allocation process. This system will more accurately identify the costs of various products and processes than the current system. Pooling to the cells is the first stage of the allocation process. The objective is to apportion and/or directly trace as much cost as possible to individual cells. The current allocation of factory overhead costs is to be phased out as the velocity of throughput in the cells reaches desired levels.

The second stage in the costing process is attaching cell costs to the individual units produced. The costs entering the pool appear heterogeneous but are homogeneous with respect to activity within the cell. The MicroChip accounting system is moving toward apportionment, abandoning allocation in the drive to improve the accuracy of costs assigned to product (inventoried or sold). This two-stage system is the focus of recent work conducted by Robin Cooper and Robert Kaplan at Harvard University.

Direct Charging: General Business's Response

Figure 7–2 illustrates General Business's revamping of its management accounting system. Management believes it is the first "value-added" accounting system developed in the United States. About 90 percent of the total costs incurred are targeted for direct product identification. The remaining 10 percent represents the basic pool of untraceable cost present in any large organization. General Business is moving along the path to activity costing.

"Direct charging," reflected in Figure 7–2, is a method to *recover* manufacturing overhead. It focuses on tracing overhead costs to specific products and targets non value-added items for elimination. Direct charging uses direct identification of expense categories, traces cost drivers through a survey of users

FIGURE 7–2
Direct Charging Methodology

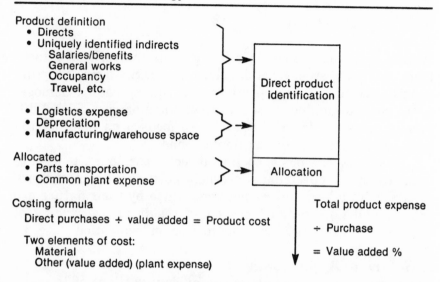

and providers, and eliminates direct labor reporting. Through these methods, the approach charges a product its actual overhead and reduces these costs based on value-added criteria. This methodology is being applied to individual manufacturing sites. The resulting measures are geared to the critical success factors for that site.

Which Costs to Trace

Figure 7–2 suggests that some items, such as travel, are uniquely identifiable to product, while parts transportation is not. These classifications, which may seem counterintuitive, represent management's best guess at one specific plant, or unit, of the company.

The costing formula suggests there are two main elements of cost: material and value-added plant expense. Adding these two subelements yields specific product costs. Dividing this total product expense by the dollar value of purchased materials provides an estimate of the value-added percentage. In devising this system, General Business does not define value-added to be solely direct labor or time/money spent on actual transformation of the product, but rather the entire set of processes and activities caused by production of the product.

Attack on Overhead

We just explored the changes being made in overhead attachment methods at several leading-edge U.S. firms. Starting from the position that apportionment (e.g., spreading homogeneous costs to product based on their common cost driver) differs significantly from allocation, we focused on defining the terms and detailing current applications.

Whenever one basis, such as direct labor hours, is used to apply the whole spectrum of overhead costs to product, the procedure is an allocation. Allocations, which by their nature are arbitrary, distort product costs and result in overpricing of high labor content products due to cross-subsidization.

A Return to Apportionment

The move back to apportioning significant portions of "pseudo-direct" costs directly to product removes the distortions resulting from arbitrary allocations. This approach, which uses the concept of a cost driver, emphasizes pooling homogeneous costs based on some key metric, or activity. As was seen at Micro-Chip, this pool is defined as the JIT manufacturing cell. General Business, whose primary concern is inventory valuation, focuses on product cost.

MicroChip is adopting two-stage overhead allocation, defined as a process where costs are first assigned to an activity center, such as the JIT cell, and then to product that flows through the cell. This allocation method reduces the number of true allocations made to product and so improves the accuracy of the final product cost numbers. The technique, found in many management or cost accounting textbooks, is finally gaining adoption more than 40 years after its development.

Where Did We Go Wrong?

We can again question the role of education in wondering why overhead procedures derailed. The process of attaching indirect costs to product has received attention at least since the time of A. Hamilton Church. The difference between allocating and apportioning is clearly defined in better management accounting texts. Two-stage overhead allocation is a traditional tool for charging the appropriate costs to the products that cause them. And *activity costing* has been in the academic literature since the early 1970s.

Management accounting is a field rich with theory, and its practitioners have put some of that theory to use. The controller at Diesel Systems, Inc., possesses a firm understanding of the basic concepts of control and how to use the numbers at his command to support management decision making. At MicroChip, Inc., the accounting group's redesigned focused-factory accounting drive addresses many of the existing shortcomings in its overhead and product costing techniques. General Business's direct charging method is apportioning the majority of its incurred costs directly to product.

Semidirect Costs: A New Concept

At each site visited, the management accounting system is being revamped to match the strategic objectives of the firm. In different ways, the management accountants at these companies are reaching into their bags of theory and tools to enhance the information they provide. There is still a long way to go before the management accounting system achieves the flexible operation needed in the rapidly changing environment of the 1990s and beyond. Never again can the system be allowed to stagnate, unchanged in the face of a changing environment. Change, rather than constancy, will be the key feature of the future.

It appears the traditional dichotomy of incurred costs into direct and indirect categories is inadequate for the proper utilization of existing management accounting theory and techniques. Throughout this chapter, the term *semi-direct* designated a third group of costs, those that can be apportioned either to an activity center or to product.

A simple addition to the vocabulary of the management accountant, this term provides a mechanism for triggering the apportionment process, for searching to find those costs that appear to be attached to some causal agent. It also embodies the move toward multiple cost pools of homogeneous cost. In attacking overhead, management accountants in all organizations are moving to regain relevance and recapture the rich theoretical basis that is their heritage.

Accountants have big trouble dealing with overhead, a black hole that swallows up everything from the equipment used to fashion a product to the security guard who watches over the plant at night. How much of the purchasing agent's salary is attributable to the semi-conductor chip, how much to the typewriter, how much to the hundred other products made in the same plant? What about the grease that keeps the machines humming, or the computers that make sure paychecks come out on time? Boiled down to its simplest form, the question becomes: Which products cause which costs?[1]

---

[1]Robert Howell and S. Soucy, "Cost Accounting in the New Manufacturing Environment," *Management Accounting,* August 1987, pp. 42–49.

CHAPTER 8

CONTROL IT WHERE YOU SPEND IT

Responsibility accounting, which seeks to prevent individuals from being penalized for outcomes they cannot control, provides a shaky basis for advanced manufacturing technologies. Holding individuals accountable for their actions does not encourage teamwork and joint problem solving in the ranks. As discussed earlier, responsibility accounting is often cited as one cause of parochialism in modern organizations.

The United States is a country built on individualism. Responsibility accounting reflects this focus on the individual. As more team-oriented approaches to manufacturing emerge, new values need to be adopted.

THE BASIC TENETS OF CONTROL

The role of control systems in organizations can be compared to the functioning of the conscious mind. Without consciousness, people do not control their own actions and do not know they are not controlling them. Similarly, when organizations cannot respond selectively to ongoing events and when they lack predesigned rules, routines, and procedures with which to respond, they have no control. Instead, these organizations are buffeted by many randomly encountered environmental forces.

Control systems are designed to keep an organization on track to ensure that what needs to be done is done.[1] Tradition-

---

[1]K. Merchant, *Control Business Organizations* (Boston: Pittman Publishing Corp.), 1985, p. 7.

ally an oppressive term reflecting the use of power and authority to force individuals to perform in some predetermined manner, *control* actually covers any effort by which an individual or organization seeks to influence the activities of another person or group. Control can be equated to motivation if properly employed, or it can be extremely oppressive if not.

Control systems differ in their use and basic philosophy. Two systems that look the same may be vastly different in their impact on behavior if their use by management varies significantly. If budgets are used as bludgeons to force individuals to respond in a lockstep manner, a defensive, rigid, unresponsive company will result. Conversely, if management control reports facilitate discussions of results and improve planning, they can provide a competitive advantage to the firm.

Responsibility Accounting and the Organization

Responsibility accounting is the accountant's contribution to the managerial control process. Six assumptions are found in most responsibility accounting systems:[2]

1. Other elements of the managerial system will be effective in inducing managers to strive to reach or surpass budgeted performance levels—the budgeted standards by themselves are not motivational.
2. The participation of subordinate managers in setting standards is necessary to the success of the managerial system.
3. Performance standards should be set at levels described as "tight but attainable."
4. The system is to facilitate management by exception.
5. Management will apply controllability criteria—that is, managers are expected to respond only to problems arising within their own jurisdictions.

---

[2]G. Shillinglaw, *Managerial Cost Accounting* (Homewood, Ill.: Richard D. Irwin, Inc., 1982), pp 691–92.

6. When the organization has two or more explicit goals, conflicts between these goals must be resolved outside the accounting system.

This set of characteristics underscores the earlier comment that responsibility accounting seems poorly suited for generating cooperative, integrated efforts across organizational and functional boundaries. In addition, the sixth point, which appears to let accounting off the hook in settling conflicts, downplays the critical role the numbers play in facilitating decisions and resolving disputes.

Controllability Fails to Foster Cooperation

Of all the above characteristics traditionally associated with responsibility accounting, *controllability* is the most likely to generate parochialism and defensive actions by managers. While it seems intuitively correct that individuals should not be held responsible for costs or outcomes they cannot completely control, operating under this concept can fragment the organization. Uncoupling performance in one area from another places arbitrary boundaries between operating and support groups.

Waste occurring in the transfer of product or activity from one group to another is not directly tracked, nor is anyone in charge of coordinating these moves in most organizations and/ or management accounting systems. The result is large buffers, waste, inefficiency, and infighting. The integrated flow of product and information so critical to advanced manufacturing technologies cannot take root in such rocky soil.

Deemphasize the Individual

Interdependence is necessary in redesigning control systems to match the characteristics of JIT and related technologies. Team output and coordinated effort are the goal, leading to decreased tracking and evaluation of individual performance. Individualism is too strong a force in American society to ignore, but it is also an inappropriate starting point for advanced technology adoption.

Interdependence surfaces in several areas. First, the clustering of machines under group technology physically stresses the relationship between a series of machines and activities. For product to flow effectively and efficiently through the cell, each production step must be smoothly linked to both preceding and subsequent activities. This highly coordinated setting isolates the bottlenecks in the process, placing a premium on eliminating these wrinkles to enhance the linearity of production.

Pull manufacturing relies on synchronized, integrated design of all activities. In many respects, the cell becomes one complex machine designed around the characteristics of the specific product family. This holistic design requires control systems that reward improved performance of the entire cell, not one individual or function. Efficiency improvements at one stage of the manufacturing process are desirable only if that area represents a constraint. Otherwise, overall performance of the cell is unchanged.

The move toward interdependence and team-based rewards cannot be avoided if the benefits of advanced manufacturing technologies are to be reaped.

PLACING THE CONTROLS WHERE NEEDED

Manufacturing cells focus on group performance, while placing little emphasis on rewarding improvements in individual efficiency. Companies using JIT cells on the plant floor should move to control systems that focus on the cell level and team output. The following examples illustrate how companies are changing their control systems to ensure that the dollars they spend translate into improved productivity and profits.

Checkbooks and Process Control

To control costs where they occur, Diesel Systems has moved to contribution reporting and adopted checkbooks as the primary element of control on the plant floor.

Checkbooks make production supervisors aware of their cost constraints. The report format is based around the production in a department. Finished products generate revenue-based deposits to the supervisor's "account." This credit is off-set by deductions for costs incurred in producing the products. The supervisor is allowed to operate the department as he sees best, given the constraint that all scheduled production must be completed on time.

Set-ups, direct materials, and labor are charged to the department. This motivates supervisors to run what makes sense (e.g., utilize the most effective and efficient machines and methods) because they want to decrease costs and inventory.

To support this effort to push control to the lowest possible level, Diesel Systems' plant manager holds a scheduling meeting with plant foremen from 6 to 7 A.M. each workday. The participation and direct control present in this system reflect a focus on responsibility accounting as the primary mechanism of control. While the supervisor is held accountable for output, it is team output, not individual performance, that counts.

Problems Exposed Up Front

The checkbook concept is maintained at the lowest level possible within the plant and is based on the total money available for the specific function. Supervisors are allowed to save for future expenses. If a supervisor exceeds the allotted budget for a certain level of output, he or she must immediately request additional funds from the plant manager. This forces a discussion of problems long before traditional cost accounting reports would be available and enhances communication in the plant.

The reports are balanced on a weekly basis, and status reports are generated for each department. This allows the supervisor to verify the specific charges and to make changes to prevent cost overruns. Accounting is using these reports to monitor its own records. The checkbook approach supports Diesel Systems' drive to switch its accounting and management philosophy from that of a job shop to a process-related just-in-time basis.

Creating Quasi Profit Centers

The use of checkbooks divides the production process into quasi profit centers. This decentralized approach to managing the plant floor provides each supervisor with the opportunity to innovate and organize to meet the unique demands of his or her subunit's product flow.

Daily production meetings integrate production and production support companywide. While this approach does not eliminate problems in the plant, the improvements in communication and performance suggest the system works well for Diesel Systems. Though triggered by the JIT manufacturing philosophy, the checkbook format yielded new procedures that are not focused solely on the departments organized into cells. The rapid feedback and visibility of the checkbook approach to management reflects the proactive, yet control-oriented, nature of Diesel Systems' management team. The company's control system modifications can be applied to JIT and non-JIT manufacturing departments. While an outgrowth of technology adoption, it is not a technology-dependent technique. Other sites are creating new accounting and control systems around their JIT systems. Capturing the essence of material velocity, zero defects, and linear production flow, these new control systems focus on team output.

Cells Are Key at MicroChip, Inc.

The focal point in MicroChip's JIT-based control system is the productivity, efficiency, and effectiveness of each cell. Performance of the cell is tracked through units scrapped, units completed, units reworked, and units for engineering. The velocity of the units through the cell is also tracked and reviewed daily.

Cell costs per hour are computed on a value-added basis. Non value-added costs are excluded in evaluating performance at the cell. The value-added components of total cell cost are: (1) direct fixed overhead, which includes direct and indirect labor, key support staff (e.g., clerical workers), and space and depreciation charges; (2) indirect fixed overhead, including production control, packaging, data-processing and training costs; (3) 68quality assurance costs; and (4) manufacturing engineering, encompassing both project costs and engineering payroll, fringes and related expenses.

Team labor is defined on a per cell basis, a unique characteristic of MicroChip's system. Not only costs are tracked at the cells. Variances from optimal production are also measured and reported.

MicroChip's accounting system also reports loss at each cell. The measure is based on the number of units scrapped times the *actual* cost at the cell. Also, yield variations and standards are no longer tracked because yield over the next year is expected to reach 99 percent. Tracking of an in-control process is a non value-added control measure.

Finally, the current plan is to transfer costs to the engineering departments for all developmental work not critical to production of existing cell-based products. The actual calculation to be used is:

$$\frac{\text{Engineering}}{\text{chargeback}} = \text{Cell time utilized} \times \text{Cell cost per hour}$$

The measure tracks development costs and monitors process time not resulting in finished products. Because cell management is to be evaluated only on value-added activities and costs, development costs must be removed from the basic calculation base.

JIT Performance Measures

TeleComm Corporation utilizes the factory-within-a-factory approach to technology implementation. This is reflected in the JIT performance measures developed to monitor cell performance:

- Good units produced/planned output.
- Cost per unit.
- Initial operability—customer service.
- Cost reduction/productivity improvements.
- Inventory turnover.
- Engineering change notices/frequency.
- Adherence to preventive maintenance and repair schedules.
- Component cost.
- Hours per machine, or total hours/units produced.
- First-pass yield of each process versus first-test yield in laboratory tests.

- Velocity of material movement (e.g., goal is currently four hours or less dock to dock).
- Daily scrap tickets.

The striking characteristic of this system of measures is the absence of controls geared toward individual performance. Each measure is team- or cell-oriented, focusing on a holistic view of the manufacturing process.

As in MicroChip's system, engineering charges are included as part of the total cell cost. The impact of engineering changes is the causal factor tracked. Performance against plan is the critical factor in evaluating actual cell output, so each of the measures either compares actual to plan or evaluates the impact of nonplanned events. Additionally, the focus on daily output reflects the basic linearity inherent in JIT manufacturing. This is in direct contrast to traditional systems that look at performance on a monthly basis or against specified jobs. Each measure rewards continuous improvement.

The performance measures required by corporate management at TeleComm reflect its preoccupation with productivity by employee and the levels of inventory (investment) maintained. These measures can be furnished by the JIT system, but do little justice to the wealth of information produced to manage the cell or to capturing the essence of JIT manufacturing.

JIT-Based Controls
TeleComm Corporation and MicroChip, Inc., gear their control systems to the individual JIT cells. Matching performance measures to the production process provides strong process control, rapid feedback, and real-time adjustment of the process. Both systems focus on the value-added features of the cell and seek to isolate and remove the impact of design, testing, and related unplanned work from the evaluation of cell performance. Tracing actual costs to the cells is feasible in this system and is done at both sites.

JIT-based control systems focus on integrating individual work units into a smoothly functioning whole. To encourage formation of teams, measures oriented toward monitoring individual performance are not used. At MicroChip, as discussed earlier, even the bonus pool is defined for team output. JIT manufacturing is triggering significant changes in the basic control system and control mentality at adopting firms.

THE CENTRALIZATION ISSUE

Controlling the money where it is spent requires a clear understanding of what basic unit is being monitored. In the factory, this can be departments, cells, or some similar subunit. The actual structure used is geared to allow the plant manager to understand, monitor, and manage the overall activities of the plant while distributing detailed decision-making authority to the appropriate level. Inappropriate decentralization of this basic authority can paralyze an organization, slow decision making, and stifle the innovation and rapid change in products and processes required to effectively compete.

Whether to centralize or decentralize decision making is a problem faced across the various levels of the organization. The decision often changes when new management comes on the scene or new objectives are pursued. An organization's degree of centralization reflects more about the management philosophy of the organization and where it has been than anything related specifically to its functioning. A major snafu in responding to a crisis can trigger a move to decentralize an organization—to add flexibility to its structure. A runaway division or plant can cause management to tighten the reins, centralizing control.

Degree of centralization is traditionally viewed as a major component of a control system. CompSci Industries and General Business, Inc., provide interesting insights into technology's impact on this age-old dilemma.

Decentralize the Accounting?
Increased use of automation, shortened product life cycles, and the movement to production management centers encompassing multiple production facilities pressure the existing man-

agement account system at General Business, Inc. Corporate management, aware of increasing distortions in the cost systems and their inadequacy for supporting management decision making, appointed a task force in 1985 to review the existing system and recommend changes.

The task force's primary finding was: "Cost accounting requirements vary from plant to plant and cannot be effectively directed from corporate headquarters." This conclusion spurred a decentralization of the costing system to match the new corporate structure and changed corporate accounting's role from a central controller to a competency center serving as an advisor and focal point for resolving cost accounting problems at individual sites.

This decentralization is especially significant because General Business, Inc., is usually visualized as an intensely bureaucratic company with a lengthy chain of command. Control in such pyramid-based organizations is usually centralized and detailed. This centralized approach hampers the flexibility and responsiveness of individual units. By redesigning the organization and its control mechanisms to match the characteristics of its markets, production processes, and geographic constraints, management is strategically positioned for continued growth and dominance in the maturing electronics industry.

Then Why Centralize?
While General Business moves away from centralization, CompSci Industries moves toward it. CompSci is traditionally viewed as a decentralized organization with redundant manufacturing facilities to support production at specific sites. Disk drive assembly provides an example of this redundancy. Many different products, or units, at CompSci Industries use a standard disk drive. In the past, each factory using these drives had its own assembly area. While this structure provided self-contained, isolated islands for innovation, the redundancy in manufacturing was very expensive. Today, faced with increased competition in its dominant markets, CompSci Industries is phasing out this redundancy.

CompSci's total performance measurement system is flexible. Since the boundaries of each level in the system are only roughly defined, the system can be easily modified to accommodate changes in the company's basic structure. It also supports more centralized management, as reflected in the basic pyramid shape adopted by CompSci management to represent this control system.

No Correct Answer

Technology does not appear to provide an answer as to why companies centralize or decentralize their basic control systems. Competitive pressures, history, and strategic demands seem to trigger changes in the degree of centralization. Since top management drives the strategic focus of the organization, it seems appropriate that strategy, not technology, is critical here.

A Return to Operational Control

Across the board, companies continue to modify their control systems to meet new demands. While technology-based measures are important at the plant level, they fade from view at the global level. In addition to centralization, or unit of analysis issues, the return to operational control prompted by advanced manufacturing technologies also represents a significant trend in the management accounting arena.

Of the many measures presented in the preceding discussion, most are not monetary in nature. In focusing on controlling the basic activities performed in a cell or a factory, cost-based measures become an outcome of the operational control system.

Upper levels of management are very concerned with financial performance—they are held accountable for it by investors. But the goal on the shop floor is to optimize the process. The process causes the costs. Managing activities, not costs, is critical at this level. A multifaceted, comprehensive management accounting/control system is formed around the strategic objectives of the firm.

INDIVIDUAL AND TEAM REWARDS: A SUGGESTION

The main issue in the basic conflict between individualism and the team-oriented technologies gaining acceptance in today's manufacturing companies is one of incentive structure design. How can a team-based reward system trigger individual excellence and growth? Because the basic fiber of American society is a belief in the freedom of each individual to pursue his or her full potential, it is a challenge to design incentive systems to promote cooperation.

The companies described in this chapter appear to be opting for a sole focus on the team, or cell. Yet, at each site, the merits of "pay for knowledge" was suggested and discussed. Combining these two subelements of the incentive system provides a possible solution to the conflict embedded in AMT adoption in American firms.

A Hybrid—Individual and Team Rewards

Figure 8–1 captures the essence of a hybrid approach to incentive structure design. The two primary components are: (1) team-based performance measures and (2) items that motivate and reward individual growth. Through participation in formal and on-the-job training programs, employees advance within their grade, or classification. This growth is based not only on their own interests, but also on designated areas for improvement in the total cell. Cross-training enhances the flexibility of the cell, allows individuals to improve both pay and task variety, and falls in line with the philosophy of continuous improvement.

At the team level, bonuses are based on actual performance against a series of indicators, all geared to ensure the continued improvement of cell-based output. Some obvious items for MicroChip, Inc., to monitor are quality, velocity of materials through the cell (e.g., cycle time), linearity, scrap, inventory turns, and on-time delivery.

Just as individual talents constrain the performance of the total cell, the team-based goals and rewards constrain the path

FIGURE 8–1
Total Bonus Pool

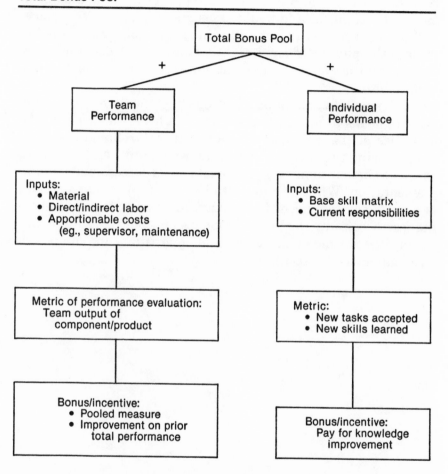

an individual can take. The two reward systems are designed to create the competition-cooperation tension often cited as a key characteristic of Japanese companies. This incentive system solution is in line with a basic belief in the value of the individual. It also recognizes that a complex manufacturing environment requires cooperation and integration across both functional areas and traditional organizational boundaries to succeed.

Triggering productive competition within an organization requires careful analysis of the unique features of that company—its strengths and weaknesses, its people, strategic goals, and basic structure. In pursuing this goal, motivation needs to replace the pursuit of rigid, tightly defined controls. Motivated people are the basis of organizational success, responsiveness, and growth.

Responsibility accounting, a basic element in the control system, needs to be adjusted from a sole focus on chopping the activities of the organization into neat, controllable packages to optimizing the performance of the whole unit. A slavish dedication to accountability and control blocks successful AMT implementation. While moving into the future, the management accounting system needs to restructure its basic tools and techniques to capture the competition-cooperation paradox. Once again, it is a time of challenge and opportunity for the proactive management accountant.

CHAPTER 9

JIT AND THE MANAGEMENT ACCOUNTING SYSTEM: PRINCIPAL FINDINGS

There is no magic button to be pushed, that one single thing which, if it were to be done, would cure the problem. An autopsy of American productivity growth would record "death by a thousand cuts" when it came to the line "cause of death." No one single thing killed American productivity growth; no one single thing can revive it. . . . Each of us is in favor of bandaging the 999 cuts caused by others and leaving the one caused by himself. Each of us envisions a patient that could survive our one cut. Each of us is right. The patient could survive any one cut, but if no cut is bandaged because each of us defends his own cut, the patient bleeds to death. And in this case the patient is our collective economic future.[1]

This thinking reflects the diversity of issues and problems manufacturers face today. How can we change our procedures to heal the cuts? In the area of the MAS, this study found pockets of change in several areas, some innovation, but inadequate resources brought to bear on these problems and challenges.

Technology Forces Accounting Change

Advanced manufacturing technologies and process modifications change the nature of doing business and require comple-

---

[1]L. Thurow, "Can America Compete in the World Economy," in *Quality, Productivity and Innovation: Strategies for Gaining Competitive Advantage*, eds. Y. K. Shetty and V. M. Buehler (New York: Elsevier Science Publ. Co., Inc., 1987), pp. 11–12.

mentary changes in accounting, control, and performance measurements. This central argument motivated this book. A traditional, rigid management accounting system, devoted to standard costing and variance analysis, cannot adequately serve the demands of the new, turbulent manufacturing environment.

Manufacturers using technology to regain, or retain, a competitive edge are doing so on a piecemeal, evolutionary basis. The management accounting systems they employ are at times lagging and at others leading the change.

Each AMT brings with it new issues and needs for information and decision-making support. The information system needs to reflect the characteristics of the technology used. These manufacturing technologies must be proactively managed and adequately monitored if effective, efficient operations are to ensue.

Islands of Change
Various technology-motivated changes to the MAS are occurring. These changes reflect the people and missions of the adopting firms. Islands of change, with evolving measurement systems, characterize AMT adoptions. While all management personnel interviewed touted their JIT lines as revolutionary, actual adoptions and change are rudimentary, encompassing only a small percentage of the companies' total productive capacity.

Our research detected a range of management philosophies, levels of technology adoption, and modifications to performance measurement systems. The majority of companies adopting JIT in their production processes are doing so on a limited basis and as a means of responding to extreme pressures threatening their competitive position.

A Lag/Lead Phenomenon
In the area of accounting change, only MicroChip's systems appear to be pulling along process improvements. At CompSci Industries and General Business, Inc., development of technology and accounting innovations appear to be occurring in tandem. Accounting/performance measures are definitely lagging

process improvements at Diesel Systems and CompSci Industries. At Diesel Systems, Inc., informal measures seem to be evolving to fit JIT characteristics, but the formal corporate system has remained rigid and traditional.

Evolving Measurements

As new measures are adopted and new technologies mastered, old measures become inappropriate or dysfunctional and should be eliminated.

- CompSci Industries and General Business, Inc., are dropping measures as they become irrelevant in the new environments.
- TeleComm Corporation's systems are virtually unchanged, except for JIT-based measures planned for the new facility.
- Both Diesel Systems and MicroChip have modified some measures. The new measurement set at these two firms includes all the former key indicators and new performance measures in certain areas.

As each of these companies modifies and updates production technology and the related performance measurement systems, outdated measures need to be discarded. The MAS of tomorrow will be significantly different, both in focus and content, from those of today. Through evolution, rather than revolution, the necessary changes are being made.

GOAL: SIMPLIFIED ACCOUNTING

Management accounting in a JIT environment is moving away from transaction-intensive data collection and variance analysis to a simpler system that focuses on cost drivers and more direct process controls. Given the simplicity of the actual process and the increased viability of process controls in a JIT environment, simplicity should be a primary characteristic of the measurement systems in these plants.

Actual changes in the field are mixed. CompSci Industries, the only firm dedicated solely to JIT, reached maturity in this process and its measurement: simplicity is its main theme. At General Business, simplicity has been embedded in product design and assembly, but complexity characterizes its automation. Simplicity in measures has not been pursued. Diesel Systems has a simple but informal set of JIT measures but is maintaining a traditional and complex MAS.

MicroChip, Inc., is in transition. The initial seven measures that serve as a focus for its accounting pilot appear to simplify the MAS, but there is no indication former complex systems are being dropped. Finally, TeleComm Corporation's complex corporate reporting system is to be forgone in the JIT shop.

Complexity causes non–value-added costs to increase. Complex measures are an impediment in a JIT environment. The collection and processing of irrelevant information cannot add value to the firm. Also, complicated measures can trigger undesirable forms of behavior. Simplicity in both operations and measurement provides the means to regain a competitive edge.

Abandon the Cotton-Ball Method

Management accountants are changing the way they pursue their tasks. Robert Kaplan, in a recent presentation, suggested it is "better to be approximately right than precisely wrong" in costing products and providing management with data. The era of the green eyeshade is departing as low-cost computers assume data-collection tasks.

We believe the management accountant should focus on developing simple, responsive performance measurement systems that support planning as well as control. This enhanced role is seen in CompSci Industries' 80 percent/20 percent trade-off between bean counting and analysis and is reflected in General Business's new accounting system.

How will the accountant use this increased capability to handle data? Will he or she continue to pursue the "cotton-ball" method, a data-intensive, complex reporting approach, or will approximations (e.g., the maxims of materiality and simplicity, currently tagged *value-added*) reshape the management accounting system? This is unanswerable at this stage.

Successful companies and their management teams are pursuing both paths.

Actual Costs Emerge

Actual costs replace the use of standards for management control in the JIT environment because of increased traceability of costs and simplification of the entire manufacturing process. Firms substitute traditional standard costing systems with systems based on actuals as they implement JIT processes and controls.

Rolling Averages Replace Standards

A new form of standard is evolving at these sites: standards are becoming a rolling average of prior actuals rather than engineered estimates. This indicates that some benchmark is needed to support performance evaluation in any setting, but engineered standards and traditional variance analysis may not be the best tools for this job.

As management accounting evolved, the information-processing constraints that hampered development of actual cost-based accounting approaches were responsible for the proliferation of standard cost-based accounting systems. As information-related costs decrease and computing power increases, it is natural that a more realistic and accurate costing system should evolve.

Actuals Support Continuous Improvement

The move from standards to actuals requires a philosophical change in goals. The goal in a standard costing system is to meet standard. Stagnation, rather than growth, is promoted by a focus on standards. But the primary goal in a JIT environment is continuous improvement through elimination of waste.

A preset standard, developed by staff engineers, does not encourage innovation and change. Instead, it rewards and supports mediocrity and gamesmanship. Meeting standard, a concept intended to increase productivity, becomes the goal of the production process. Overhead is not reduced, it is absorbed, building, rather than eliminating, inventory. In a perpetual quest to avoid reporting variances and attracting management attention, production managers and employees lose sight of the actual goals of the organization and instead channel productive energies into data manipulation.

Changing the Evaluation System

Evaluation of floor personnel is beginning to reflect the movement toward actual costs. JIT techniques appear to be moving control closer to the point of production and to be more clearly matching responsibility with controllability of costs. For example, Diesel Systems instituted a checkbook for cost control within the work center that focuses solely on direct costs (e.g., traceable). Also, the company is revamping its internal reports to isolate overhead controllable only at the plant level as a lump-sum figure in a contribution margin-based reporting system.

At TeleComm Corporation, managers in the JIT shop will be evaluated solely on the new measures—actual costs and linear production levels. CompSci Industries is using the migration in measurements to evaluate employee performance. The pilot JIT accounting system at MicroChip, Inc., reflects a focus on actual unit costs. Because of problems noted earlier, the new system will be a hybrid of standards and actuals. Also, final evaluation of inventory will continue to be based on standard costs.

At General Business, Inc., the new direct-charging system is eliminating standards and substituting rolling averages. Only overhead remains as a traditional allocation, and it will probably be revamped as the new system matures. At each site, new technologies affect the basic measures being used to both manage operations and drive the performance evaluation process.

Actuals—Just a Starting Point

Although abandonment of the traditional standard accounting system is not obvious, most of the sites studied appear to be adopting an actual cost system. Given the costs of maintaining an information system and the confusion caused by the existence of multiple data sources, this multiplicity is probably only a temporary phenomenon. The movement toward actuals is hampered by corporate controllers and management's continuing demand for the traditional standards-based reports. Until top management becomes comfortable with using actual costs, individual plants and divisions will find it difficult to adopt sweeping accounting reforms.

The rolling-average-of-actuals standard being adopted by the firms in this study coupled with reward systems that encourage improvement and penalize lack of improvement (see MicroChip) provide an alternative approach to motivating desired behavior on the plant floor.

Simple substitution of actuals for standards is not enough. People are being asked to live with ever-tightening standards and goals on the production floor. To do so, they must know the reasons for this goal, accept it, and believe they are not merely pawns in a game. Eliminating waste needs to become a companywide goal. Merely squeezing waste out of production areas will not solve the problems facing manufacturers. Replacing people with machines will simply change the type of noise top management hears.

Isolating Non Value-Adding Costs

Isolating non value-added costs focuses management's attention on these items. The objective is to move these "burden rate" generators to zero. Eliminating non value-added costs requires measurement of key characteristics of JIT improvements, such as defects, cycle time reductions, and obsolescence. Figure 9–1 details the progress being made in measuring these items.

Each research site is measuring and reducing/eliminating defect rates and cycle times. Obsolescence is a different matter. Little progress is being made to adapt existing MAS measures.

Overhead: Some Improvement
Diesel Systems' internal reporting eliminates uncontrollable overhead from its plant performance system. It focuses on over-

FIGURE 9–1
Tracking Non Value-Added Costs

Company	*Costs*				
	Operations-Based Non–Value-Added	*Manufacturing Overhead*	*Obsolescence*	*Defects*	*Cycle Time*
Diesel Systems	Measured	Measured	Recognized	Measured and reduced	Measured and reduced
TeleComm	Measured	Measured	Not recognized	Measured	Measured
CompSci	Measured and minimized	Measured and reduced	Recognized	Measured and minimized	Measured and reduced
MicroChip	Measured and reduced (by cell)	Recognized	Not recognized	Recognized	Measured and reduced
General Business	Measured and reduced	Measured and reduced	Recognized	Measured and minimized	Measured and minimized

head as a lump-sum cost to be addressed by plant management rather than as an allowed, hidden part of product costs.[2] MicroChip, Inc., is segregating value-added from non value-added overhead, although allocations are still being employed. While total elimination of non value-added activities is an unattainable goal because of mandated disclosures, regulation, and taxation issues, reduction is being pursued.

Continual Improvement: A Goal

Value-added characteristics are measured as part of the elimination of nonprocessing time at the work cell level. This was observed at all the research sites. The flexibility and respon-

---

[2]In reviewing the results of this study, Professor Thomas Johnson raised the concern that management at Diesel Systems is confusing strategic product cost information with operational control. Prior research would suggest that all overhead should be traced and applied to products for strategic planning support, but only controllable costs should be included when measuring the cost and performance of manufacturing.

siveness of the new technologies allow companies to achieve massive reductions in throughput time.

The potential savings depend on more than a capital equipment acquisition decision. The need to minimize cycle time—to keep materials moving because the only time value can be added to a product is when it is being processed—must be recognized. Materials waiting for processing create costs without resulting in any improvement in the product. Cycle time in this setting consists of four key elements:

$$\text{Total cycle time} = \text{Process time} + \text{Move time} + \text{Wait time} + \text{Inspection time}$$

Only processing time adds value; the rest add cost. Reducing total cycle time means reducing cost and increasing competitiveness. Improving and maintaining total cycle time requires development of cycle time-based measures. One example of a measure of manufacturing cycle efficiency (MCE) is:[3]

$$\text{MCE} = \frac{\text{Processing time}}{\text{Processing time} + \text{Wait time} + \text{Move time}}$$

This measure captures the essence of the goal of optimized manufacturing: eliminate waste. This elimination recognizes that all costs are not equal. Value-added criteria provide a firm basis for analyzing the activities of the organization, resulting in targeted cost reductions and eliminations that enhance the firm's competitiveness.

The level of integration and coordination of performance an organization attains within the total performance measurement hierarchy determines whether the company achieves manufacturing excellence. Isolated changes may provide momentary relief from competitive pressures, but only by adopting the philosophy of continuous improvement across functions and manufacturing settings can excellence be attained and maintained.

---

[3]*CAM-I Report,* 1987, chap. 1, p. 5.

SUMMARIZING MAS CHANGES

Each company modified its MAS to replace the traditional standard cost system with one based on actuals and/or rolling averages of actuals. The systems observed run the gamut of complexity, and many of the modified systems are a hybrid of old and new methodologies. Other than the focus on actuals, the changes being made are heterogeneous.

Diesel Systems is pursuing a contribution margin approach geared toward control. It is being maintained simultaneously with a corporatewide standard cost system. These new reports are informal but are seen by plant management as being more relevant than the old. TeleComm Corporation is in a state of flux, while CompSci Industries formally changed its performance measurements to reflect the four criteria for excellence: people, quality, cost, and delivery. Simplicity is being pursued, but the treatment of overhead at CompSci Industries remains fairly traditional. At MicroChip, Inc., the most interesting change made to the MAS is the formal measurement of cycle time and cycle time-based performance measures. Finally, General Business, Inc., is decentralizing its MAS and adopting a direct-charging approach that will attempt to more accurately trace costs to products and processes.

Pockets of Change

Combining all these changes into one system would be a definite step forward for the MAS. The helter-skelter adoptions noted suggest the MAS is in an early stage of transition. Additionally, it appears that one consistent solution will not arise. Instead, it is likely the MAS will become a specifically tailored management tool that supports the strategic focus of the individual firms. This variety in methods and approaches reflects the various organizations' cultures, strategies, product and process mix, and information needs of managers. Given the MAS objective to serve management decision making, this heterogeneity of systems seems to be a step in the right direction.

Proactive Management of Change

Adoption of JIT technology implies a proactive management style. Proactive change entails planning and recognizing the costs and benefits of the alternative manufacturing processes. The change is based on a commonsense approach to management and the automation of only those processes and procedures that are under control and that provide for progress on the four criteria of manufacturing excellence: people, cost, quality, and delivery.

This suggests MAS changes should be based on a desire to match the characteristics of the MAS to the different technologies. Reactive decision making based on competitive pressures underlies most technology adoptions. The basic control and support systems are not being mastered before technology adoption (e.g., leapfrogging is occurring), and, for the most part, accounting changes are lagging technology adoptions.

The bottom line is that companies must actively manage change in their products and processes if they are to become world class manufacturers. Adopting a manufacturing strategy that will lead to excellence and a competitive edge requires proactive decision making and planning. It also requires that these activities maximize the efficiency and effectiveness of the manufacturing process and eliminate waste and activities that do not add value to the company or the final product.

Translating the strategic goals of the organization into the performance measurement system provides management with a means to manage change and channel employee behavior. Proactive management suggests that changing measurements and incentives are critical for successful JIT adoption.

JIT Measures: New Insights

Diesel Systems is relying on informal systems for JIT measures; TeleComm Corporation and MicropChip, Inc., are piloting JIT accounting procedures; and CompSci Industries and

General Business have reached maturity on this dimension. Only General Business has made significant progress beyond the JIT level.

Two key variables influence the clear-cut ranking implied by this analysis: (1) matching the characteristics of the measurement system to those of the process used by the firm and (2) the related complexity of the product mix produced at the specific sites.

These results, coupled with implementation problems, suggest two specific conclusions:

1. When firms attempt to leapfrog technologies, or, more specifically, move to new technologies before mastering the lessons of previous approaches, it increases the probability that the accounting systems will be unable to support the demands of the new systems, and hence will be perceived as a hindrance.

2. When firms attempt leapfrogging, accounting systems that initially lag this jump can be utilized to enforce the organizational learning inherent in the skipped technologies.

The Future: It Depends

The advance of AMTs addresses more than resource consumption. If the major task performed by most management accounting departments is inventory valuation, the concept of zero inventory poses serious implications for the survival of the profession. But if the MAS expands to fill its other, neglected roles, this trend will prove to be a boon rather than a curse. Increased awareness among practicing managers and accountants that MAS numbers drive behavior would be one of the most important events in the history of the managed firm.

DIRECTIONS FOR THE FUTURE

Eli Goldratt regrets ever having called cost accounting the number one enemy of productivity . . . because everybody is jumping on the bandwagon condemning cost accounting when we should be condemning cost, not accounting.[4]

While this statement might appear to let the MAS off the hook, the chapter's opening comments warn of the dangers of this type of approach. If the management accountant believes the damage caused to the organization by tracking incurred costs rather than the cause of those costs can be left unbandaged, he or she is contributing to the decline and possible death of manufacturing in the United States. The patient's pulse is already weak; further negligence at any level in the organization could prove fatal.

The major changes needed to update the MAS can be summarized as:

- Cost drivers need to be identified.
- Accountants need to expand beyond traditional financial reporting roles and activities.
- Simplicity of measures and methods needs to become the overriding goal of manufacturing.
- The MAS approach chosen needs to be tailored to the specific environment or company.
- Direct labor tracking and reporting needs to be deemphasized.
- Pilot approaches should be used before major system changes are undertaken.
- Large-scale concept education of both financial and nonfinancial management needs to occur.
- Work orders need to be eliminated or minimized.
- The goal should be to so minimize raw material and in-process inventory levels that one account, and a reduced number of transactions, can be used to control these areas.
- Overhead allocation methods need to be reexamined and changed to match the features of the production process.

---

[4]S. Jayson, "Goldratt and Fox: Revolutionizing the Factory Floor," *Management Accounting*, May 1987, p. 18.

The MAS must focus on cost drivers and use them to identify and eliminate non value-added costs. Only when the causal relationship between costs and the processes and procedures that generate them is clearly understood can management intelligently eliminate non value-added components. Without this understanding, any attempts at cost reduction has great potential for harm. Most managers can cite cases where across-the-board cost-reduction projects have increased cost, rather than saved their company money.

Each time a cost driver is identified and utilized, the information generated by the management accounting system increases in relevance. Use of these cleanly defined activity measures eliminates the distortions arising from arbitrary allocations.

Expanding Roles
The second recommendation in the above list captures the essence of the types of change needed—not replacing the accountant with a computer but rather expanding the role to encompass more than product cost. CompSci Industries indicates what these new roles will be.

Control is a necessary evil in organizations. Control mechanisms alone cannot move a company to excellence because they are oriented toward the past. While tracking operations is essential, motivation and improvement are necessary to succeed in the competitive environment of the future. In a JIT setting, planning and coordination are as critical as control. In most cases, the processes have self-contained control mechanisms. The successful management accountant in the future will need to assume a proactive position on the management team.

Strive for Simplicity
Simplicity has been suggested as one way to minimize non value-added costs and enhance a firm's competitive position. As noted earlier, two leading manufacturing firms take different approaches to this concept. Both have simplified their basic processes. CompSci Industries is pushing simplicity to the limit in its accounting systems and relying on the strengths of JIT procedures to maximize its competitiveness.

Alternatively, General Business, Inc., is putting complex technologies and complex accounting systems into place—automating the JIT processes. Intuitively, the simple process would appear to have the greatest flexibility with lowest costs. In this case, hindsight will be the best way to compare the relative merit of the two strategies.

Tailor the MAS

Each organization and manufacturing facility is a unique environment. General Business believes this is so, and has decentralized its management accounting system. The types of changes made to the MAS will need to be tailored to the specific environments.

This belief stems from the concept of cost drivers and their central role in the MAS of the future. These drivers will vary from site to site as will the strategic objectives, culture, and structure of the organizations. No one uniform system can serve this heterogeneous audience.

Deemphasize Direct Labor Tracking

Direct labor tracking and reporting need to be deemphasized in the future MAS. As technology is installed and the labor component of product costs decreases, the costs of labor reporting rapidly exceed the benefits gained from the information gathered. To ensure that the information system serves a value-added function, costly measures with little or no information value need to be identified and eliminated.

In many companies, this change has already occurred. Rather than focusing on incurred costs, the management accountant must begin to more adequately trace costs to products and processes. It is essential for management accountants to help their companies eliminate those costs that provide no enhancement to the firm or its products and profits.

Attack Overhead

New allocation methods need to be implemented that will correspond more closely to the activity, or driver, leading to the specific cost. Some firms are beginning to avoid allocations that violate the controllability criterion inherent in responsibility accounting. Others are using multiple activity measures as the basis for the allocation process.

If an overhead cost is truly value-adding, it should be open to market solutions that pit the quality, responsiveness, and cost of the internal service to market prices for similar services. Some firms are using this approach, allowing intrapreneurship to replace centralized control as the basis for maximizing operations of subunits. This solution is surely one way to beat back overhead creep. Another is an increased focus on value-added concepts at all levels of the organization. The final objective in both cases is the same—elimination of waste.

Rely on Pilot Systems

Pilot approaches may be the best way for individual companies to modify their MAS to meet corporate objectives and subunit needs. Islands of accounting change are occurring. This piecemeal, but carefully planned, evolution in performance measures and the role played by the MAS in the organization is desirable.

These islands represent innovation and competitive strength. Over time, companies may want to migrate their entire business to the island and then abandon the old, leaky ship they were using. The island is the final goal, but before it is adopted totally, the evidence must bear out its ability to support life over the long term.

> The appearance of these new articles and cases may signal that cost management is returning to U.S. manufacturing firms for the first time in more than 60 years. The(y) . . . describe concrete procedures for compiling cost information that serves managers, not accountants. They show how to garner cost information that helps managers identify and evaluate the resources needed to deliver value to the customer. These new systems do not focus managers' attention on accounting costs but identify and focus attention on transactions and events that consume resources. The power of these new systems rests on the principle that people cannot manage costs—they can only manage what causes costs. That is what cost management is about. Information for cost accounting has nothing to do with it.[5]

---

[5]Johnson, 1987, p. 12.

CHAPTER 10

JIT ACCOUNTING: A LOOK TO THE FUTURE

The management accountant can learn to think "just-in-time" accounting to keep in step with desirable changes in management style. It is not possible to re-do a whole accounting system in one quantum leap. The management accountant must adjust, however, to changing technology and new management philosophies and be willing to meet the changing demands of information. Nothing is more discouraging than to hear financial people say, "We can't accommodate that change because of our accounting system." Accountants would do well to become familiar with the latest management philosophies and advances in technology so that they can help their firms stay in step and just-in-time as well.[1]

Competitive pressures among American manufacturers over the past 20 years are changing the structure of industry. These evolutionary changes are being adopted within plants, divisions, and corporations on a piecemeal basis, creating chaos in our current reporting, measurement, and control systems.

Manufacturers currently appear to be functioning with three reporting methodologies, which are both formal and informal in their usage and composition: Process control, managerial reporting (MAS), and financial reporting. Each one services the needs of a unique constituency. Process, the lowest

---

[1]R. Seglund and S. Ibarreche, "New Challenges for the Management Accountant," *Management Accounting*, August 1984, p. 38–46.

level, serves the production supervisor; managerial, plant management and other internal users; and financial, the outside interests, either corporate or the financial markets. As suggested by Figure 10–1, these reporting methodologies can place conflicting demands on the basic information system.

Designing three separate, complex reporting systems to meet the diverse needs of these constituencies is inconsistent with competitive requirements. Alternatively, performance reporting and accounting can be approached in a hierarchical manner, defining at each level specific elements to be reported and controlled while utilizing one comprehensive database. This basic concept is embedded in CompSci Industries' performance measurement system. The following sections detail one conceptualization of a hierarchical MAS, starting with a description of the role played by competitive objectives, a dis-

FIGURE 10–1
Information System Demands*

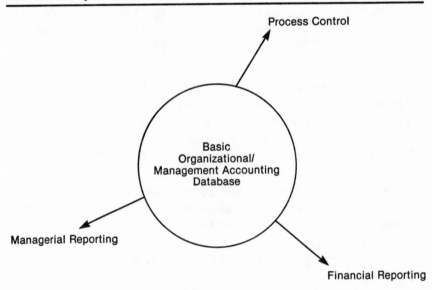

*But - Can one information system support these conflicting demands?

cussion of the underlying theorem, and finally one feasible control methodology that reflects this view of the MAS.

Competitive Objectives

The first step in developing a reporting hierarchy for an individual business unit is to define the competitive objectives. In the manufacturing sector, these goals should be consistent with, and supportive of, those for the organization as a whole. A tripartite planning structure, such as that suggested in Figure 10–2, is one way to achieve the necessary integration.

FIGURE 10–2
A Tripartite Structure

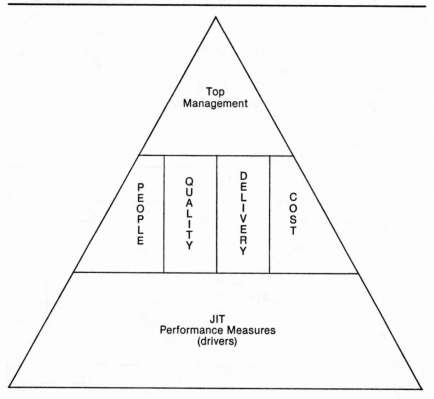

The manufacturing objectives can be summarized into four counterbalancing areas that reflect the criteria for excellence discussed previously:

1. Quality.
2. Cost.
3. Delivery.
4. People.

Figure 10–3 depicts this type of approach. The overriding objectives are defined in broad terms at the top of the management hierarchy based on current performance versus planned performance on each of the four criteria for excellence. These goals are then broken down into subgoals, making sure each level in the managerial hierarchy is evaluated based on the counterbalanced, complete set of objectives. In the manufacturing area, tools such as JIT can be used to embody these strategic objectives.

Simplicity Is the Starting Point
The key ingredient in identifying these four areas is the simplicity of their counterbalancing natures. The pursuit of these objectives can be clearly measured, providing the organization

FIGURE 10–3
A Total Performance Approach

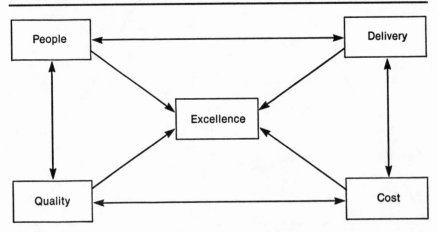

with specific directions to guide its actions and decision making.

As various pressures arise, the short-term needs of the organization can be balanced by strategic objectives. The measurement system should signal when these situations occur. For example, a drive to meet sales or delivery objectives at the expense of quality, cost, or people should be clearly signalled through the management structure via the MAS.

Make the Accounting Reflect Strategy

Construction of the measurement hierarchy must clearly reflect the strategy of the enterprise. While this discussion assumes equality among the four areas to signal a situation (i.e., when sales takes precedence) requiring management attention, many enterprises have little explicit equality in these areas. What is critical is that: (1) the MAS embody the strategic objectives of the enterprise, and (2) the MAS clearly and quickly signal any occurrences that deviate from the planned path. Through various weighting and reporting schemes, the company can focus on specific aspects of the overall strategy of quality, cost, delivery, and people. All four need to be represented and monitored by the MAS—each is critical for the long-term health and growth of the organization.

MINIMIZING INTERVAL

One overriding principle guides the structure of these objectives. The goal of a manufacturing enterprise should be to:

MINIMIZE INTERVAL.

Current activity in leading-edge companies focuses on interval reduction, commonly referred to as just-in-time. JIT activity is concerned with the manufacturing process on the shop floor—both in terms of quality and basic material flows. Additional areas targeted for improvement are procurement, materials management, and performance measurement.

These steps already show substantial innovation in many mature JIT locations. However, they lag behind movements confined to the shop floor. These procedures and principles must be expanded beyond the plant floor to encompass the entire organization.

JIT is not merely another inventory technique nor a way to manage work flow in a manufacturing plant. JIT and the other manufacturing environments discussed earlier require cultural and strategic changes within the adopting organizations. To attain excellence and a competitive edge, all levels and all functional areas must address the elimination of waste.

The Core Concept

The underlying concept that will drive the merger of the three existing information systems is:

$$COMPLEXITY = COST$$

$$SIMPLICITY = SOLUTIONS$$

Applying the above principles will ensure the most efficient accounting and measurement hierarchy, moving from conflicting information systems to a JIT-oriented conversion of the MAS. Simplicity is embedded in JIT and is the critical element in design for manufacturability in a CIM environment. Complexity creates the need for coordination and integration mechanisms, which are predominantly non value-added cost drivers. Simple methods and simple structures provide a means to identify and eliminate fat from the organization. They are essential for firms wishing to succeed in the turbulent manufacturing environment of the 1990s.

JIT ACCOUNTING: AN APPLICATION OF THE CONCEPTUAL FRAMEWORK

The first stage of the MAS design process is the identification, observation, and recording of key activities, or cost drivers, for the process or product.

Characteristics that influence cost in a JIT environment include the complexity of the product, characteristics of the cell, existing capacity, and effectiveness of the production cell. The complexity of product that drives cost include (1) number of material moves, (2) number of total parts, (3) average number of options, (4) number of total products, and (5) number of engineering changes.

Cell characteristics that can be used to evaluate the cost-causing potential of the cell are the hours each machine is run, the first-pass yield of each process in the cell, the velocity (e.g., cycle time) of material moving through the cell, and the number of direct labor employees. As the first-pass yield decreases, the cost of the cell increases because errors cause a pause in production. If material moves slowly through the cell, more cost is incurred. The key for each of these metrics is to continuously strive to push the cell performance toward its optimum as a holistic, unified production cell, much as one would optimize process forms of manufacturing through balanced flows of materials.

Cost drivers emerge in existing capacity. Set-up time and number required both limit the actual capacity of a plant. As these factors increase, the cost per finished unit increases. If the cell is run without attending to preventive maintenance schedules, then costly breakdowns can occur. On the quality front, statistical quality control (SQC) capabilities limit the effectiveness of the JIT cell. As the number of process changes increase, cost in the cell rises because of learning curve and variety effects.

The metrics used to meter cell effectiveness do capture cost-causing elements of performance. Examples are: (1) good units produced per planned output, (2) daily scrap levels, or number of tickets (e.g., number of times scrap was detected and line stopped for correction), (3) relative cost of components, or material, per total cost of product, and (4) number of units reworked or scrapped later in the production process. Each of these metrics represents ineffective elements of the manufacturing process and can be assigned the appropriate cost.

Through these and other measures of cost drivers, the JIT accounting cell provides a means to understand how various changes in the production process affect the total cost of the products made in those cells. In addition, the JIT accounting cell lays the foundation for an improved system of control that more accurately identifies the area of performance hindering continued improvement. Turning now to more traditional classifications of the components of manufacturing, what types of measures can be developed in a JIT accounting system to meter material, labor, overhead, and the benchmarks used in evaluating improvement?

Material Reporting

In the area of materials, the most significant change to the MAS in a JIT setting is replacing raw materials and work-in-process inventories with a streamlined raw-in-process approach. This basic change eliminates many transactions from the MAS.

The manufacturing ideals found in JIT techniques support this restructuring of inventories because all materials are delivered straight to the manufacturing floor. This reduces move and queue time, and hence cost.

Other changes to the MAS include eliminating all purchase orders using blanket contracts and direct communication between vendor and user to integrate the supply chain. Overhead becomes a direct charge to specific products and processes, utilizing increased traceability and a focus on eliminating waste as the impetus for employing Church's theories on overhead reduction and apportionment.

Finally, all volume-related performance measures, standards, and variances need to be restructured and/or eliminated. In a setting where the goal is to approach linear production and to build only to order, not to stock, traditional measurements that reward unneeded production must be eliminated from the MAS.

JIT and Labor Reporting

In the area of labor reporting, operator standards and time/attendance standards need to be eliminated. The move to get rid of direct labor reporting, such as that undertaken by CompSci Industries, reflects this drive to simplify labor reporting and refocus management attention on more costly aspects of the production process.

Three other major, labor-based modifications to the current MAS are dictated in a JIT setting: (1) measure performance based on hours per unit produced, (2) employ exception vouchering in place of variance analysis, and (3) utilize a team manning approach (see MicroChip, Inc.). The first two points provide an alternative to the detailed, standard costing approach to labor reporting. Relative efficiency can be assessed simply by summing the total number of hours used in a specific area and dividing it by the units produced during a certain period. This productivity measure captures the relationship between inputs and outputs in a value-free manner that allows employees to focus on improvement rather than attainment of a preset standard.

Exception vouchering is a viable alternative to variance reporting and analysis in the labor area. Rather than waiting for end-of-month reports to detail problems on the plant floor, plant management gets immediate feedback from production. Additionally, defective materials and related problems can be detected and rapidly removed from the production process. Management becomes part of the problem-solving mechanism, communication between levels is enhanced, and the MAS is transformed from a mechanism for control to a communication device and a tool for coordination.

Control in a JIT Shop

Team manning approaches to labor reporting in the MAS and the use of cell-based cost pools reflect an essential difference between the new and the traditional manufacturing environment. A summary of factory cost reporting in a JIT setting is shown in the boxed feature presented at the top of the next page.

JIT Factory Cost Reporting

Cost collection elements
   Team labor: Production, maintenance, quality assurance
   Die and P.P. charged at standard cost through B.O.M.
   Equipment depreciation
   Maintenance and quality assurance support
   Power usage
   Space cost
   Other operating expenses
   Administration/engineering, etc.

Cell cost per hour of operation
   Compute and track monthly

Collect costs of each cell into an indirect department
   Move to actual cost system; no labor or overhead standards
   Stop direct labor reporting as currently done

Variances at cells
   Purchase price variation to piece part standards
   Value-added variations
      Spending (budget versus actual)
      Lost time (actual lost time × standard cell cost/
         hour) capacity/volume measure
      Velocity (planned versus actual time at actual cell
         cost/hour) efficiency/speed

Value-added costs at cells
   Cost of good units = process theoretical time (not actual time
      in cell) × cell cost/hour

Non–value-added costs
   Costs − allocation between inventory and period expense
      Variances
         Spending (budget versus actual)

Report scrap loss at each cell
   Number of units scrapped × actual cost at the cell (Die and
      P.P. is at standard)
   Stop reporting yield variation; no yield standards

Cost transfer to engineering departments
   Cell time utilized × cost per hour

The key element in this cost reporting system is the cell. All costs can be traced to the cell. This also means that individual performance, once the focus of the reward and penalty system in the plant, is now impossible to separate from team (e.g., cell) output. In a JIT line, such as those utilized by CompSci Industries, individuals downstream from a production stoppage are unable to work. Total production is performed by a team of workers that is only as strong as its weakest link.

Because the team functions as a unit, it should be measured and rewarded as a unit. MicroChip is moving toward this approach in its pilot accounting model. Additionally, team manning helps eliminate the troublesome, arbitrary differentiation of direct and indirect labor in a JIT setting. If technicians are essential for efficient production, they are part of the labor pool in the area, as are the machine tenders.

This change to the MAS will reflect reality and reduce distortions. Each of the labor-based changes suggested above simplifies the reporting process and therefore incorporates the message contained in the overriding theorem of management in a JIT environment.

Overhead: Friend or Foe

The most important message in the list of overhead tracking changes is the attack on overhead creep through a value-added versus non–value-added approach. Efficient, effective use of indirect functions in the organization provides the means to regain a competitive edge.

As voiced by Church, every cost is caused by a process. Direct charging based on traceability criteria may replace the standard costing approach, which seeks to apportion these costs rather than control or reduce them.

Allocated overhead costs are basically hidden and unquestioned. The focus in a traditional setting is producing enough to absorb all the overhead. The financial accounting-based need to neatly categorize and record each dollar of resources consumed results in a vicious cycle of adjustments, whose purpose is to eliminate variances, a non–value-added concept, rather than improve performance.

The message is simple: minimize allocations and question those that remain. Each cost incurred has a cause—the MAS needs to be reoriented from an ex post scorecard of incurred costs to a proactive, performance improvement mechanism.

Finally, in planning a new product or process, management should assess its impact on support area costs. Neglecting this has two consequences: (1) project costs are understated, and (2) overhead charges increase overall. Failure to recognize the impact of changes in process on the costs incurred in support areas distorts product costs, can result in wrong decisions, and short-circuits attempts to identify value-added costs and combat overhead creep.

Standards: Eliminate or Streamline

A review of the types of changes organizations are undertaking suggests a far-reaching redefinition and reevaluation of this basic component of the MAS. The increased reliance on actual costs in the firms studied here has been underscored. Firms are increasingly replacing standard costs with actual costs and compiling rolling averages of actuals to serve as a benchmark for monitoring performance and improvement. Instead of comparing actual cost to engineered standards, it should be compared to a benchmark average of historical performance.

Transactions Cause Cost

A second major change discovered in the field is the move to increasingly shallow bills of materials. Transactions cause cost because information is accumulated, integrated, and distributed based on these transactions. More transactions translate to increased information-processing demands within the organization, resulting in large costs incurred merely to coordinate and record activities. Table 10–1 illustrates the changes a JIT approach to overhead charging can make in final product costs.

Combining backflushing with the two changes discussed above creates a simplified product cost tracking system that increases the velocity of information flow as the number of repetitive, low information-value transactions is reduced. Back-

TABLE 10–1
Commercial Load

Item	Old Method	JIT Method
Purchasing	$100	$ 50
Production control	100	30
Receiving	100	10
Stores	100	10
Material handling	100	10
Inspection	100	0
Commercial engineering	100	100
Total	$700	$210

flushing of materials does not require standards; actual usage and cost is available from the MAS. In total, the emerging MAS promises to be a system dedicated to processing the minimum number of transactions needed to support decision making and provide a performance report on process efficiency.

Life-Cycle Costs Are Important

Closing the loop in process and product management requires management of product life cycle-based costs. Increasing competitive pressures demand optimum performance from an organization at all stages in a product's life. And the traditional period for recouping development costs is shrinking rapidly.

In this turbulent environment, development costs must be recouped shortly after production begins. As the time from product launch to maturity compresses, successful manufacturers will enter the market with competitive costs that quickly secure market share and profits. Paving the way for these achievements, CAD technology provides a firm with the ability to design for manufacturability, improve process design, and ensure quality enhancement.

Cycle Time Variances Emerge

Cycle time variances replace volume-based standards in the modified MAS of the future. Volume-based approaches encourage unneeded production because they tend to base performance evaluation on overhead absorption (e.g., minimize volume variance) rather than on effectiveness and efficiency of the production process.

Cycle time measures focus on reducing non–value-added costs, such as move and queue time. Their objective is to measure improvement toward the goal of minimizing interval. Cycle time measures support the movement to optimized manufacturing and encourage change and improvement rather than defensive reporting and evaluation.

Impact of JIT Accounting

JIT accounting entails a tremendous number of major changes in the policies, procedures, and systems used to design and operate the MAS. A brief list of these changes includes:

- Accounts receivable matching system will automatically match purchase orders to the appropriate receiving documents and invoices.
- Materials and related "A" item costs will be backflushed rather than tracked on a detailed basis.
- Incoming inspection requirements will be drastically reduced, as qualified vendors' materials are delivered directly to the production line.
- Tracking of work-in-process inventory will be eliminated, replaced by a new raw-in-process approach.
- Work orders will be eliminated.

This list of changes holds three primary trends: (1) JIT costing approximates process accounting, (2) actuals replace standards as the primary basis for applying costs, and (3) an increased focus on traceability replaces volume-based costs.

JIT approximates process accounting mainly because of the use of backflushing to relieve inventory. In process accounting, equivalent units of production, determined at the end of the period, are used to charge cost back to work-in-process versus finished goods inventory accounts. In JIT accounting, the transfer of cost is based on completion of the production process, at which point the various inventory accounts are reconciled, manufacturing labor is attached to the output, and suppliers are paid for materials used. Relatedly, shop floor reporting is done at gateways, again providing the basis for reconciliation in much the same manner as in process accounting. As in process accounting, the integrity of the JIT accounting system depends on the integrity of the bill of materials, the actual yield obtained on the raw materials used, and the ability to detect product quality problems early.

In the second trend, JIT accounting utilizes actuals rather than engineered standards in evaluating performance. Flexible standards based on average cycle time are used to measure performance against schedule, rather than a preset goal. Since these benchmarks are constantly revised (e.g., they are a rolling average of actuals), a new variance is created: a standard revision variance. If this period's performance exceeds historical average, the performance variance is positive. Conversely, negative values indicate cell effectiveness has deteriorated. The goal is to always have a positive variance.

Finally, JIT accounting replaces volume-based costs with an emphasis on traceability: direct versus indirect costs. Within a JIT cell, many previously indirect costs can be assigned directly to the cell. These assigned costs may be fixed or variable—the key is that their incurrence can be traced to the specific cell. With those indirect costs that remain (e.g., overhead), accumulation should be done by cell and assignment by machine hour or cycle time metrics.

Units of production depreciation will begin to reign in this setting, as the driving issue becomes, "What causes these costs? How can I minimize the distortions in tracing indirect costs to product?" If cells are evaluated based on cycle time, not total volume of output, then machine cost should be matched as closely as possible to time used. Units of production depreciation reflect this goal. Finally, life-cycle reporting is supported in JIT accounting because the costs of learning a new process are built into the measurement system. Early in a new product or process run, costs will be high because cycle time is high. As the learning curve takes over, these costs decrease and so does the accounting system's measure of them. On the plant floor, at least, the cost measures evolving in JIT settings support life-cycle cost objectives.

THE MIGRATION OF
MEASUREMENT SYSTEMS

An enhanced role for the management accountant entails developing a total performance measurement system that is flexible enough to provide necessary operating information for various levels of managers. Such a system is a far cry from the systems of today, but the picture of the demands on the MAS of tomorrow is still incomplete. The concept of dynamic change, of a migration of measurement systems to match the technologies used by the firm, is still missing.

Migrating through Measurements

A migration appears in the performance indicators needed in the various environments. Items critical for supporting a basic manufacturing process are subsumed by an MAS used to track performance in an MRP or JIT environment. As CIM is reached, the number of measures utilized decreases significantly.

The continuous flow of materials indicated success in earlier technologies, but a CIM environment is built on the concept that time is money. The technology itself reflects the basic process-related control measures. Maximum performance is achieved on design and manufacturing procedures, and efficiency and effectiveness are captured by the total time elapsed from the inception of a new product until it is delivered to the customer's hands. Minimizing total cycle time maximizes the profits of the organization while optimizing performance, as judged against the criteria for excellence.

Monitoring Design Improvements
The JIT environment reduces the number of measures used. As layout, process, material handling, and related functions are optimized by proper application of the method, they no longer need to be monitored by the performance measurement system. The simplicity of the JIT setting and the pursuit of standardization remove the need to track the impact of variety and multiple product options on the efficiency of the production process.

The final step to CIM results in almost total elimination of performance measures for the design process. CAD/CAE procedures optimize the design process. The key criteria for gauging performance become lead time to design a finished product and startup time needed to bring a new product from design to production. Only time-oriented measures remain, as computerized systems are employed to maximize the efficiency of the design process as well as the actual configuration of the final product.

Measures of Material-Related Performance

The same approach can be used to examine the types of changes that will occur in materials-related performance measures as an organization moves through the various technologies. Unlike the design area, a basic control system embodies many of the cost drivers in materials management. Component lead times and customer service levels are the exceptions but are entered into the MAS once MRP II is properly implemented.

In a JIT setting, the measures used decrease rather than increase in number. Because of the one-day window for production and the underlying philosophy dictating that each component arrive as it is needed, inventory stocking policies become irrelevant. Material handling, WIP inventory turnover ratios, and storeroom stocking locations measures suffer the same fate. The entire process is tracked by simple observation. The MAS in a CIM environment is slightly simpler but approximates that used under JIT manufacturing.

Measuring Production-Based Performance

A basic control system is capable of capturing all the key dimensions of the production process except those related to the value-added concept. Within an MRP II system, these neglected measures are added. As with the design measures, they are the most comprehensive at the MRP II stage.

Two JIT trends simplify the production measurement system: (1) total quality control and (2) zero set-up times.

Production linearity becomes an important measure of efficiency in a JIT plant and remains as a key production performance measure.[2] Also, value-added measures remain, as the drive to eliminate waste continues.

A CIM environment includes only the value added per employee and the actual value of the produced output. Lead time disappears as cycle time approaches optimum. Non value-added costs are eliminated in JIT and so need not be measured in a CIM plant. Lot size becomes irrelevant because one of the objectives of JIT manufacturing is to drive feasible lot sizes to one. The measures currently seen as crucial for monitoring the performance of the production process are eliminated, as computerized systems and improved procedures employ common-sense methods to eliminate waste, not monitor it.

Cost Accounting in a JIT Environment: Implications for the Future

Basic changes occurring in manufacturing techniques and technologies amplify existing shortcomings in the MAS and provide the needed impetus for changing that system. Current practice became obsolete not through mass ignorance or criminal negligence by the management accounting profession, but through an incomplete understanding of the complex role played by this information system in reflecting, recording, and creating organizational reality. The linkages between product costing and financial reporting, variance analysis and performance evaluation, and standards/overhead allocations to budgeting and strategic planning suggest that modifications to this corporate language (e.g., the MAS) are not to be undertaken lightly.

Technology Has Served to Outmode the MAS

As currently structured, the MAS was designed for an environment with few products and few work centers. The traditional full-cost accounting methods that applied standards by

---

[2]Linearity means, simply, producing the same number of units each day. Based on projected sales, or booked orders, needed production is divided into equal "buckets" for daily scheduling. This simplifies the maintenance of raw materials and adds predictability to the system.

products and variances by department or work center were an adequate information basis. Information-processing constraints underscored the need to maintain one information system. The financial accounting-oriented system that evolved was probably the most cost-efficient tool for tracking product costs and meeting the demands of external reporting.

Current problems can be traced to increased complexity in the manufacturing environment, which now supports many products passing through multiple work centers and departments. Increased complexity requires improved and expanded information flows. Global variances by functions and work centers cannot be relied on to track complex causal factors that drive costs.

Pursuit of simplicity in process isolated the distortions created by a mismatched MAS. When the production environment was undisciplined and unnecessarily complex, the specific weaknesses in the information system (MAS) were hard to pin down. The long-standing frictions between production and accounting indicate that these distortions in the accounting system are not new.

The loss of relevance in the MAS was gradual. Repairing it cannot not be gradual because the rapidly changing environment of the future will increasingly emphasize the relevance, speed, and relative accuracy of performance information. To meet these demands, the MAS must look to simplicity in its accountings to enhance its capacity to process information.

Critical Success Factors in a JIT-Based MAS
Visibility and rapid feedback will become the bywords of a JIT system, matching the characteristics of the MAS to the characteristics of the technology.

Variance reduction will not be focused on performance against engineered standards, but will instead focus on variance-causing aspects of the production process: quality, total cycle time, and the number of parts. This reflects the importance of production linearity in a JIT setting. The MAS of the future will need to be migratory and flexible, sensitive to the changing demands of the market and management, as it moves to trace costs to the activities that cause them.

Internal Control Is Still Important

The changes we suggest will have significant impacts on the auditability and the strength of the internal control process in adopting firms. Moving to paperless purchasing procedures, eliminating WIP inventories, modifying depreciation methods to a units-of-production approach, changing overhead allocation schemes, and building in constant change could be a nightmare for the auditor.

Yet, the conceptual framework and measures suggested here provide more timely, accurate information than exists today. A redefinition of *audit trail* and related concepts likely will evolve from the redesign of the MAS. Audit concerns should not hinder MAS redesign, nor can they be ignored. Education of the accounting profession will need to expand beyond the boundaries of the firm.

Move JIT beyond the Plant Floor

The future of manufacturing competitiveness lies in the rapid development of JIT and other AMT principles beyond the shop floor. As labor and material costs even out throughout the world's markets, the value-added use of overhead will provide the competitive edge.

The lessons found in the concepts of interval reduction and the elimination of non–value-added activities need to be applied in research and development, engineering, accounting, order processing, distribution, and other traditional staff functions. These efforts will reduce management hierarchies and thus increase the velocity of decision making. As suggested by A. Hamilton Church, the glob called *overhead* needs to become our friend in the pursuit of efficient, effective manufacturing.

CHAPTER 11

AMT ADOPTION—MATCHING THE SOLUTION TO THE SETTING

Generating the culture of continuous improvement among all employees and other constituencies of the company is top management's responsibility. Continuous improvement is not a national or regional culture. It is a management culture, deliberately and painstakingly created by management. If upper management nurtures and protects the culture, technicians can do many marvelous things, but technicians cannot grow manufacturing excellence from a culture that rejects it.[1]

A brief review of the popular press on American competitiveness and the role of technology in generating improvements leaves the impression that many see technology as the cure for all ills. Losing market share? Get flexible—streamline your operations with CAD/CAM and JIT. Inventories out of control? Use MRP and Kanban squares to force the fat out of the system. The list goes on and on.

Technology cannot change a culture or ensure the attainment of manufacturing excellence. Excellence can thrive in an unautomated setting while remaining illusive in a CIM plant. Two critical elements are necessary to reach this goal:

1. Technologies must match the characteristics of the company and its environment in order to be used successfully.

---

[1] R. W. Hall, *Attaining Manufacturing Excellence* (Homewood, Ill.: Dow Jones-Irwin, 1987), pp. 264–65.

2. AMT adoption is more than plugging in a new machine. It requires changes in the basic culture of the company and a firm commitment of top management time and effort.

Throughout this book, JIT has been described as a journey—a philosophy of doing business that reaches beyond the plant floor to encompass every aspect of the firm, touching the daily lives of each employee. A basic cultural change, JIT manufacturing presents a challenge and an opportunity to American manufacturing firms.

MAKING THE TECHNOLOGY DECISION

Hayes and Wheelwright suggest that manufacturing strategy can be partitioned into eight basic decision categories:[2]

1. Capacity: amount, timing, and type.
2. Facilities: size, location, and specialization.
3. Technology: equipment, automation, linkages.
4. Vertical integration: direction, extent, balance.
5. Work force: skill level, wage policies, employment security.
6. Quality: defect prevention, monitoring, and intervention.
7. Production planning/material control: sourcing policies, centralization, decision rules.
8. Organization: structure, control/reward system, and the role of staff groups.

Categories 1, 2, and 3 have one thing in common: each project in these groups originates within the capital budgeting process. Capital budgeting is one of the processes subsumed under the MAS in most companies. Firms use capital budgets to allocate their scarce resources among competing projects.

Project justification is one of the most hotly discussed topics associated with the impact of advanced technologies on the

[2]R. Hayes and S. Wheelwright, *Restoring Our Competitive Edge: Competing Through Manufacturing* (New York: John Wiley & Sons, 1984).

manufacturing environment. Companies seeking to enhance their competitive position must compete on multiple product dimensions (e.g., cost as well as service and quality). To meet these new strategic demands, manufacturing firms are turning to the flexibility and promise of advanced manufacturing technologies. To date, these projects are being predominantly justified on faith alone. Managers are abandoning the cumbersome formal justification procedures encompassed by traditional capital budgeting procedures.

A Series of Weaknesses

The most notable shortcomings of the capital budgeting process with respect to the new technologies are: overreliance on quantified (e.g., cash flow) results, inadequate recognition of strategic and product improvements (e.g., qualitative) benefits, short time horizons, and unrealistic hurdle rates. Changing these shortcomings does not require abandoning the capital budgeting process, but it does demand a clearer understanding of the strengths and weaknesses of various components. Qualitative objectives can be easily used to augment quantitative analysis. Hurdle rates can be adjusted to more adequately reflect the true cost of capital. Through these minor adjustments, capital budgeting becomes once more a managerial tool that helps enhance the value of the firm.

Exploring Capital Budgeting and AMTs

We have known for a long time that there is no one right way to analyze a proposed capital investment. . . . The availability of . . . information transforms the capital-investment analysis from opinion into diagnosis, that is, into the rational weighing of alternative assumptions. Then the information transforms the capital-investment decision from an opportunistic financial decision governed by the numbers into a business decision based on the probability of alternative strategic assumptions. . . . What was once a budget exercise becomes an analysis of policy.[3]

---

[3]P. Drucker, "The Coming of the New Organization," *Harvard Business Review,* January 1988, p. 46.

The analysis of capital investments has long been equated with financial techniques such as return on investment (ROI) and payback. Depicted in the literature as a decision-facilitating tool, capital budgeting issues have been seen as relatively unproblematic.

Project justification is one of the weakest areas for the existing MAS. The underlying objective of capital budgeting techniques is to support the intelligent management of the firm's investments in capital assets. Yet, numerous studies show that capital budgeting serves more as a basis for rationalizing management intuition than for a priori evaluation of the true merits of different projects.

Current questioning of project justification techniques raises two issues:

1. Are the techniques no longer capable of supporting good management decisions?
2. Are the techniques still appropriate, but most AMTs fail to provide benefits that exceed their costs?

Both of these observations probably have an aspect of truth. The qualitative, long-term benefits, such as enhanced quality and responsiveness, of many AMTs are not currently incorporated in the capital budgeting model. This shortcoming can be remedied by using various modeling and survey approaches to redesigning the project justification process (see CAM-I Report, 1987, for one such approach).[4]

The second problem is not a weakness in the analytical techniques but rather in the decision-making process. CIM is not for everyone, at least not in the foreseeable future. The discipline of MRP and the simplicity of JIT need to be conquered before a firm is ready to jump to more advanced technologies. Leapfrogging technologies is not only hard to cost justify, but it is also not good business.

Improve Postimplementation Audits
The only way to fix the shortcomings of the justification process is through improved postimplementation audit procedures of

---

[4]*CAM-I Report,* 1987.

FIGURE 11–1
Cost of Quality

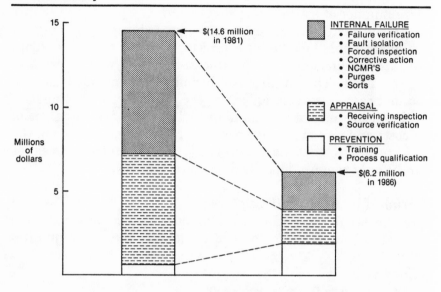

existing projects. Figure 11–1 presents one such followup, as performed by Xerox Corporation. This graph documents the significant cost savings achieved by improving quality through total quality control techniques coupled with JIT.

These savings were probably almost impossible to quantify into cash flows before the fact, but ex post the savings are evident. A way is needed to factor management intuition into the justification process in a formal manner. This objective can be achieved, but at the same time we must prevent adoption of technologies for the wrong reasons or as a perceived quick fix to deep-rooted manufacturing problems.

To Date: No Change

The companies studied made no obvious changes to project justification procedures. They appear to be abandoning traditional techniques where necessary and moving forward on intuition in these situations. Project justification remains an unresolved problem. Managers are adopting JIT and other AMTs based primarily on intuition and faith. They are abandoning their traditional capital budgeting techniques where possible or completing the necessary forms to meet corporate goals even if the cash flows used are basically fabricated on a large number of estimates.

Traditional project justification procedures focus only on the quantifiable benefits of a project, so firms trying to adopt advanced technologies face a difficult situation, as suggested in Figure 11–2. The qualitative benefits not included in discounted cash flow techniques, for the most part, will translate in the future into quantifiable impacts on the firm in terms of its costs and its competitive position. Now, representation of the opportunity costs is lacking. These forgone future revenues can be only roughly identified by management.

Rational or Rationalizing?

Are the techniques of traditional justification obsolete? Can they, or should they, be modified to simplify the use of qualita-

FIGURE 11–2
Qualitative versus Quantitative Benefits

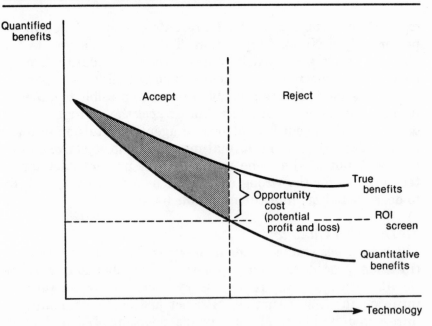

• Projects not passing the ROI screen may cause opportunity costs!

tive factors? This discussion focuses the "true" role played by project justification procedures.

Research indicates that actual cash flow analysis is usually undertaken to rationalize investment decisions that have already been made. The underlying problem is that technologies such as JIT are difficult to rationalize. They are strategic moves that cannot, and perhaps should not, be rationally tied to discounted cash flow techniques.

Modifying existing techniques is simple if the only true question is enhancing the role of qualitative indicators. The fact that change is not occurring suggests that alternative explanations exist that have not yet been detected. This research does not shed significant light on this issue. The strategic issue in the adoption of CIM by General Business, as compared to continued devotion to JIT by CompSci Industries, is one area exposed by this study. But more research needs to be done before any conclusions can be reached.

Managing Numbers Is Not Managing a Business

No progress in the area of project justification has been discovered, and a multitude of people are asking why. Perhaps it is because we are addressing the wrong issues or in the wrong manner. In a society where a numbers-oriented management style has prevailed, the messages in the new technologies run deep. Managers and their organizations are changing as they face competitive crises that threaten their existence. This study maintains that this process of change must be proactive.

Managers must once again take an active, risk-taking, innovative approach to their jobs. This message of change is embedded in JIT and related technologies and makes their justification so difficult to rationalize.

Use Different Justification Techniques

Discussions with experts in the management accounting field suggest another twist on the project justification issue.[5] Much

---

[5]These insights were originally suggested by Pat Romano of the NAA during meetings conducted to discuss the results of this research.

like the A-B-C approach to inventory management, a ranking of magnitude and strategic import can be applied to a class of capital assets and/or projects in the justification process. To assume that such projects are the same is a critical flaw in the analytic, cash flow-oriented capital budgeting techniques currently in use.

There will probably always be a set of projects, such as a simple die replacement, best judged on the basis of a hurdle rate. There always has been and will be a group of projects that so greatly affect the strategy or culture of an organization that quantitative methods of analysis will falter.

These tools are not inherently good or bad; their use dictates their final impact on organizations and people. Different classes of projects will undoubtedly need a new set of analytic tools. We need to enhance the bag of tools as well as recognize their limitations if progress is to be made in capital budgeting.

Project Justification

We believe strategically oriented projects may need well-informed management intuition to get off the ground. JIT requires a change in a company's approach to doing business. The cultural and strategic changes it entails cannot and should not be relegated to a set of cash flows.

This approach to capital asset justification reveals the role played by risk in any human endeavor. Analytic techniques can reduce this uncertainty and provide information to aid decision making. But in the end it is managers, and their intuition, which make or break an organization.

> A lot of management uses a capital budgeting model like a drunk uses a lamppost—for support, rather than illumination. A capital budgeting technique should allow you to illuminate the various possibilities. It should allow you to ask where we should put our money and what we should do with it, as opposed to just asking how we work to get around this, to work around this particular area.[6]

---

[6]D. Dilts (with G. Russell), "Are Accountants Delaying the Automation of America?" *Cost Accounting for the 1990s, Proceedings, National Association of Accountants,* 1986, p. 34.

UNIQUE SETTINGS FOR AMT ADOPTION

The adoption of technology represents a unique challenge for every company. It affects the organization's culture, modifying its basic fabric. Deal and Kennedy, in discussing corporate culture and change, suggest:

> Culture is the barrier to change . . . it causes organizational inertia; it's the brake that resists change because this is precisely what it should do . . . protect the organization from willy-nilly responses to fads and short-term fluctuations. Yet . . . change is often necessary.[7]

Diffusion is the process by which an innovation is communicated among members of a social system. It is the impetus for cultural change, as old ties and practices are discarded and new ones undertaken in adopting and implementing the innovation. Hawley discusses this point:

> Man reacts to the web of life as a cultural animal rather than as a biological species. Each acquisition of a new technique or a new use for an old technique, regardless of the source of the origin, alters man's relations with the organisms about him.[8]

The business of change in organizations is cultural transformation.

Change Is Not Accomplished Overnight

The process of diffusion is slow. The overall time elapsed in the technology adoption (e.g., diffusion) process can be affected by: (1) urgency—in times of real crisis people understand the need to change, listen to suggestions, and adapt more quickly; (2) attractiveness of the proposed change to the affected individuals; and (3) strength of the culture being changed.

Strong, ingrained cultures are very difficult to change, often exhibiting what has been called the dysfunctions of bu-

---

[7] T. Deal and A. Kennedy, *Corporate Cultures: The Rites and Rituals of Corporate Life* (Reading, Mass.: Addison-Wesley Publishing Corp., 1982), p. 159.

[8] A. Hawley, *Human Ecology: A Theory of Community Structure* (New York: The Ronald Press, 1950).

reaucracy—the inability to change due to entrenched structures and subgroups. This type of problem is common to all the sites visited. Only TeleComm Corporation is having difficulty breaking away from its cultural blinders, though. Rogers details five major factors that affect the extent, and speed, of diffusion in an organization:[9]

1. Relative advantage of old versus new.
2. Compatibility, or the degree to which an innovation is perceived as being consistent with the existing values, past experiences, and needs of potential adopters.
3. Complexity of use and, relatedly, understanding of the innovation.
4. Trialability, or the degree to which the innovation may be experimented with on a trial basis.
5. Observability of results to others.

Cultures, or companies, do change under pressure, and the degree and rate of change is determined by the relative impact of the new methods on the existing culture. Also, risk factors for the individuals involved play a major role in limiting grand-scale change.

The various advanced manufacturing technologies do not represent equivalent types of changes for the adopting organizations. JIT can be adopted in a piecemeal fashion but is a high-visibility move requiring significant changes to the existing mindset. CAD/CAM is a less radical process to adopt. JIT adoptions lag CAD/CAM adoptions. Finally, CIM is not feasible on a piecemeal basis and requires more adaptation across the board. The earlier migration path to excellence can be thought of as a series of innovations that lead to greater amounts of cultural and procedural change in the firms that adopt them.

Stages in Diffusion

There are two primary stages in the diffusion process: initiation and implementation. Initiation covers agenda setting, in-

---

[9]E. Rogers, *Diffusion of Innovations,* third edition (New York: The Free Press, 1983).

formation gathering, and matching the innovation to an identified organizational problem. It is done before making the adoption decision.

Following adoption, the implementation process and the real problems for the organization begin. Implementation consists of redefining/restructuring the innovation as well as the organization. Clarifying and routinizing are also part of the diffusion process. During this sequence of events a dual change occurs: the innovation is modified to fit the organization, and the structure of the organization is changed to accommodate the innovation.

Make the Technology Fit Your Organization

The diffusion of an innovation can be seen as a source of learning for an organization, requiring varying degrees of change to its underlying social structure, or culture. The level of change required directly affects the speed, extent, and success of the diffusion process. Change in each organization is unique.

No two cultures, or organizations, are alike. The actual application of JIT technologies will look and feel different at each company, and the accounting techniques needed to support decision making in the firms will differ significantly. No cookbook answer, taken from the shelf, can provide a quick fix to competitive problems. Both the technology and the information system that supports it need to be molded by active top management involvement and attention to the unique needs and characteristics of the company.

Management Philosophy: Providing an Explanation

A second explanation for both the technology and performance measurement system hierarchies presented in this book lies in the concept of organizational cultures, or management philosophy. While the benefits of technology, such as product improvements, flexibility, and responsiveness, are promoted by articles and books in the popular press, the role of management in shaping the path of the organization is virtually ignored.

An alternative premise arising from this discussion is that management philosophy and/or acceptance of the JIT philosophy is critical to its successful implementation.

We observed that as top management's acceptance of the basic JIT philosophy increased, so did the speed of implementation of JIT procedures and the usage of JIT concepts in non-JIT areas of the company. To the extent that acceptance lags the implementation of a technology, problems can occur, both recognized and hidden. A brief list includes:

1. Management dissonance can create roadblocks to AMT success.
2. The benefits available from the focused production inherent in AMTs with regard to logical physical flows, such as decreased cycle times and inventory levels, will not be fully exploited.
3. Vendor control (e.g., partnership), a critical factor in successful JIT implementation, may not be adequately pursued.
4. Accounting and control methods may remain relatively unchanged, hence management persists in a reactive rather than proactive style.

Two basic forces can be diagrammed for each firm, as suggested by Figure 11–3. This figure illustrates the maxim, "JIT is not a technique, it is a journey—a philosophy of doing business." This simple idea contains a complex interaction between management philosophy and management practice. Chris Argyris has labeled this problem as the difference between "espoused theory" and "theory-in-use": a glamorous way to redepict a phrase known well by all parents with adolescent children, "Do what I say, not what I do."

An organization is a group of people working toward goals. Where those goals are set and how they are transmitted through the organization are questions for future research. What is of importance here is the gap between management's stated objectives and its true beliefs, motivations, and criteria in establishing corporate policy.

If not supported by a strong commitment to the goals of simplicity and visibility, a JIT line will fail to reach its full potential. We believe the benefits inherent in JIT are so over-

FIGURE 11–3
Management Philosophy and JIT Implementation

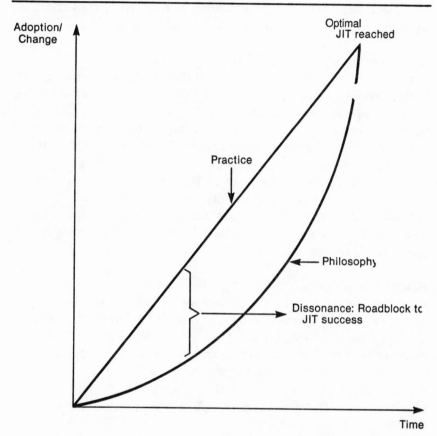

whelming that even halfhearted acceptance of the methodology will reap large benefits for the adopting organization. Perhaps one of the unsung merits of JIT is that it can, like an evergreen, take in even poor soil. While its growth may be stunted, its survival is ensured in all but the most hostile environments.

An Increasing Awareness of the Impact of the MAS on Behavior

One final issue in a JIT environment is the increasing awareness of the impact of the MAS on behavior. Since the early

1950s, research has documented such phenomena as goal displacement, the dysfunctional consequences of control, management myopia, and a host of other behavioral problems that arise from the use of numbers to evaluate people.[10]

Numbers Suggest Hard Facts

People attach more credibility to numbers than words in analyzing performance or making decisions. Because of their objective appearance, numbers are seen as truth. Yet most accountants are aware that accounting is more of an art than a science. This characteristic of accounting provides the discretion needed to support management decision making.

Most of the shortcomings of the MAS can be ultimately tied to the impact of numbers on behavior—the mystical shroud that surrounds the accountant. Because these fabricated symbols (e.g., product costs) form the basis for behavior, the noted flaws in the MAS need to be addressed.

Accounting Numbers Change Behavior

If accounting numbers were used as a guide in decision making, there would be room for error in their calculation. Instead, they are used to evaluate decisions: individuals are fired because of reported performance, and projects are accepted or rejected on the basis of cash flows generated by the accounting system. While the management accountant may wish to avoid this responsibility for behavior, it is an inherent part of the function he or she performs.

The impact of recorded versus unrecorded transactions may provide one explanation for why sunk costs continue to be used in making investment decisions while opportunity costs are ignored. Sunk costs represent past decisions, transactions that are recorded and reported by the accounting system on a periodic basis. They are a living reminder of expended resources, and their presence on management reports underlines their importance. Sound arguments have been made by academics and practitioners alike that opportunity costs are a

---

[10]Some representative works are Ridgeway, 1956; Argyris, 1953; Hofstede, 1967; and, Hopwood, 1972, 1976, 1986.

critical aspect of any decision.[11] Yet the manager reviewing the period reports sees no mention of opportunity costs. Not recorded or reported must mean unimportant for decision making.

Management Accounting Is Not Power-Neutral

Management accounting is inextricably tied to behavior in organizations. Its use in performance evaluation alone suggests any changes to this system must incorporate the behavioral impact of the new techniques. Accountants and managers have virtually ignored the motivational impact of information generated by the MAS in their current drive to adapt the MAS to meet the demands of the new technologies. Changes such as those undertaken at MicroChip, Inc., have far-reaching implications for employee behavior. Rather than improving performance, they may decrease it. This all encompassing role for the MAS needs to be acknowledged and addressed. The MAS is not a power-neutral analytical tool. It both reflects and constructs organizational reality.

The movement to change the MAS is far more significant than developing a few new measures or changing the basis of allocation. American manufacturers are discovering that, although they are prolific producers of goods, they are not as efficient or effective as the Japanese. Focusing solely on economies of scale, manufacturers have paid little attention to the costs of coordinating these massive enterprises or to the impact of increasing uncertainty in the environment on the effectiveness of these rigid structures.

Regaining the Leading Edge

Information system demands are based on the complexity, uncertainty, and interdependence that exists both within an organization and between the organization and its environment. These demands have changed significantly over the past 20 years. JIT is highlighting this change.

---

[11]See Shillinglaw, 1982, for an excellent discussion of opportunity costs and their use in analyzing business decisions.

Why have MAS changes lagged? Why has the MAS failed to warn and prepare managers of the need to change technologies and organizational structures? The answers to these questions may be mass ignorance, but they may equally be because the MAS served as a basis for control and power in the organization. And it has served these purposes well, so there has been no demand for change. Today, managers are asking for scanning and planning information in the search to decrease uncertainty and provide the basis for strategic decision making. The types of changes being suggested for the MAS reflect this shift.

As technology adoptions proliferate, MAS changes will occur. The solutions that emerge will not look or feel Japanese. They will be the American versions of the philosophy of continuous improvement, modified to reflect the physical realities of production in the United States. Japan is small—the United States is massive. Suppliers are clustered close to manufacturers in Japan and solely dedicated to the parent company. In the United States, free competition rules suggest that market diversification is the best way to ensure ongoing success. In fact, government regulation impedes development of dedicated vendor networks. Added to these basic differences is the purely logistical problem of moving materials over the distances that often separate the manufacturer from its supply network.

Improvements can and should be made to existing manufacturing processes. These improvements will be unique to each firm and setting. Solutions that succeed in one firm or market may destroy another company. Because of this, technology adoption needs to be carefully planned and executed. Top management needs to be involved to guide the change process and be ready to champion the cause of continuous improvement and the regaining of a competitive advantage based on defined product attributes, such as cost and quality, rather than size or tradition.

> An organization is a drawing—varying, changing and dependent very much upon the ideas of the people who create it. Since accounting is part of the organization, it should be treated as one of the strokes of the pen that make up the drawing.
>
> Steve Fleckenstein, 1987

Appendix A

RESEARCH METHODOLOGY

> In behavioral research, naturalness or naturalism . . . is a function of what the investigator does . . . the set of activities an investigator actually engages in while conducting his research falls somewhere in a two-dimensional space. The first dimension, which is most frequently thought of in differentiating research activities, describes the degree of the investigator's influence upon, or manipulation of, the antecedent conditions of the behavior studied, on the assumption that the degree of such influence or manipulation may vary from high to low, or from much to none. The second dimension, which is less commonly considered than the first, describes the degree to which units are imposed by the investigator upon the behavior studied.
>
> —Williams and Raush, 1969, 46.

Naturalistic research focuses on becoming part of the context of the organization/problem being studied in order to search for meaning, a meaning which is representative of organizational reality as a whole rather than fragmented characteristics of the firm. It is a research approach based on the concepts of "grounded theory" [Glaser and Strauss, 1964]. The key factor driving this methodology is its recognition that the "researcher is forced into the natural setting because he or she cannot specify, without an a priori theory, what is important to control or even to study" [Lincoln and Guba, 1985, p. 43]. It is a qualitative methodology which utilizes inductive data analysis, emergent design, negotiated outcomes, and a case study reporting mode to build theory.

One of the primary concerns in undertaking research is the validity of the findings—their trustworthiness. This is defined on four criteria [Lincoln and Guba, 1985, p. 290]: "Truth value," applicability, consistency, and neutrality. The in-depth discussion of the merits of naturalistic inquiry on these criteria can be found in the underlying text cited here. It is a research method directly opposed to traditional positivistic research and one that espouses the merits of organizational context in understanding the interrelationships of variables and constructs in impacting behavior in organizations.

The research approach used in this study is a direct reflection of the characteristics noted above. The research propositions evolved from the information gathered during the first site visit as well as from the prior experiences of the researchers. The theory itself emerged from analysis, both analytic and intuitive, of the research settings and characteristics of the organizations and individuals studied.

The details of the objectives, research areas, and actual on-site schedule are reproduced in depth following this introduction. The semistructured interview document used is also reproduced in this appendix.

The research team consisted of three individuals (the authors). For each site visit, at least two of the three researchers were present to conduct the interviews. A semistructured interview format was used to allow the collection of as much data as possible regarding the unique characteristics and problems of the individual firms. The sample consisted of five manufacturing firms drawn from Coopers & Lybrand's client list, representing a fairly broad and random spectrum of technology adoption and MAS redesign.

Each site visit was designed around a three-day format. The actual length of each visit varied based on the characteristics of the firm, the availability of management for interviews, and the constraints placed by the participating firms on the total amount of data they were willing to disclose. While the format used was flexible, the areas noted in the original research objectives was adhered to. Hence, an emergent, inductive, and grounded approach was utilized to collect the data for this study.

I. PROJECT OVERVIEW (NAA CASE STUDY)

Objective

Prepare five manufacturing case studies dealing with the issues of cost accounting and financial reporting in a Just-in-Time/Advanced Manufacturing Technology (JIT/AMT) environment.

Approach

1. Select five candidate manufacturing sites to be studied.
2. Interview key management personnel in the following areas:

    General Manager,

    Engineering,

    Materials Management,

    Production,

    Finance.

3. Analyze and summarize interviews and client documents into case-study format.

The Cost Accounting Issue

Just-in-Time is a philosophy dedicated to the elimination of waste. The JIT philosophy strives for simplicity, minimal lead time, and value-adding functions throughout operations. While the objectives of the philosophy can be simply stated, the changes necessary to achieve the objectives are often profound. JIT often results in the implementation of process-flow redesign, synchronization of production to design, small lot sizes, and pull-system scheduling. The cumulative impact of these activities changes the nature of doing business and by definition requires complementary changes in accounting, control, and performance measurement. Efficiency and detailed work-center cost measurement, the underpinnings of many existing cost accounting systems, would likely be eliminated in systems that support JIT. Accordingly, the redesign of cost accounting systems will occur on an accelerating basis as more and more manufacturers make the philosophical change.

In order to effect these changes, serious hurdles must be overcome. What system changes were made? What were the changes in accounting policy? How were internal controls impacted? Within these broad accounting-systems issues, we will focus on cost accounting issues such as:

- Product justification,
- Performance measurements,
- Cost behavior patterns during the product life cycle,
- Value-added and non–value-added costs,
- Treatment of direct labor,
- Allocation techniques,
- Depreciation methodologies,
- Definition of cost centers (Group Technology),
- Definition of fixed, variable, direct, and indirect costs,
- Inventory valuation,
- Development of standards for forecasting.

Product Justification
In addition to quantifiable costs such as direct labor and direct material, to what extent are factors such as quality, flexibility, throughput time, customer responsiveness, and market share used to justify new technology? How are projects related to stated business objectives?

Performance Measurements
How are the factors that were used in the project justification measured? (Lower costs, improved quality, shortened production cycles, greater responsiveness to changing requirements, lower inventories, etc.)

Cost Behavior Patterns during the Project Life Cycle
As products move through their life cycle, the significant tasks of manufacturing change. How are these tasks identified and measured?

Value-Added and Non–value-Added Costs
What methods are used for classifying costs as either value added or non–value added? How are the various classifications of non–value-added costs presented to management?

Treatment of Direct Labor
How is direct labor defined, measured, collected, and allocated? What other costs are based upon direct labor information?

Allocation Techniques
What methods are used to define and choose a basis of allocation that most accurately associates the expense with a specific cost objective? How are cost centers related to overhead costs to products?

Depreciation Methodologies
What are the various methods used for calculating depreciation and what are the guidelines for selecting a particular method? As the manufacturing environment changes, does the allocation method change?

Definition of Cost Centers
Careful attention to the definition of cost centers is required to ensure recording the appropriate data at the correct level of detail. Who is responsible for defining cost centers and what guidelines are used?

Definition of Fixed, Variable, Direct, and Indirect Costs
What are the guidelines for classifying each element and how are they used in developing standards and reporting to management?

Inventory Valuation
How are inventory carrying costs calculated and reported? How is the impact of reduced lead time and improved throughput measured?

Development of Standards for Forecasting
In order to control costs at the level at which they are generated, it will be necessary to predict the costs that will be incurred at various activity levels. How are the significant levels of activity identified and how are the costs developed and selected?

Objectives of the Study

The objectives of the proposed research study would be to:

1. Document the overall JIT philosophy with particular regard to accounting issues that surface during a JIT implementation.
2. Select and document five case studies which portray how the selected companies dealt with the accounting issues. What systems changes were made? What were the changes in accounting policy? How were internal controls impacted?
3. Document the findings of the research in a format suitable for NAA distribution to its membership.

II. PROPOSED INTERVIEW SCHEDULE (NAA CASE STUDY)

	Estimated Time
Day 1 A.M.	
General/plant manager and key staff, introduction and project explanation	½ hour
Financial manager, reporting system overview	1 hour
Plant tour	2 hours
Day 1 P.M.	
Production manager, performance measurements	1 hour

	Estimated Time
Engineering manager, performance measurements	1 hour

Day 2 A.M.

Materials manager, performance measurements	1 hour
Cost accounting manager, cost accounting system detail	3 hours

Day 2 P.M.

General plant manager, business plan/manufacturing strategy	1 hour

Day 3 A.M.

Financial manager, detailed discussion of issues	3 hours

Day 3 P.M.

General manager and key staff, overview of issues and project findings	2 hours

APPENDIX B

QUESTIONNAIRE

COMPANY: _____   DATE: _____

Plant Manager

I. Business Overview   *Response*

1. What is your formal business plan?
2. What is the manufacturing strategy/objectives of your operation?
3. What method/procedure do you use to justify investment opportunities?
4. What performance measures are currently in place?
5. How do you identify cost drivers?
6. What performance measures do you use?

II. Management and Support Organization Structure   *Response*

1. How many organizational layers exist in your company?
2. What is the basic organizational structure?
3. What degree of autonomy do you have with respect to decision making?
4. Is overall control in the organization distributed or centralized?

5. Approximately how accurate is your forecast?
6. How often do you change/update your forecast?
7. What external factors affect the way your organization/area functions?

III. Performance Measures/PLC   *Response*

1. What type of performance measures are in place?
2. What are your process capabilities/limitations?
3. What is management's philosophy regarding life cycle costs?
4. What kind of information is available regarding these costs?
5. How accurate is this information?
6. How much product variability do you experience?
7. What is a typical life cycle for your products?

General Comments:

Production Manager

I. Production Overview   *Response*

1. Do you produce every piece according to demand?
2. How accurate are your forecasts with respect to actual customer demand?
3. Who generates your production forecasts?
4. What is your plant's capacity?
5. What is your manufacturing philosophy?
6. What type of manufacturing policies are in place?

II. Plant Layout/Lot Sizing   *Response*

1. What is the general layout of your facility?
2. What is your routing procedure?
3. How complex is this procedure with respect to your present layout?
4. How diverse is your product line?
5. How many part numbers do you utilize?
6. Do you have many common/similar parts with different part numbers (degree of standardization)?
7. What is a typical lot size for your production?
8. What lot sizing techniques do you use to determine lot sizes?
9. How do you measure capacity utilization?

III. Setup/Cycle Times   *Response*

1. What are the setup times on your machines?
2. What percent of the setup is external? Internal?
3. What is the cycle time of your products?
4. How many schedule changes do you typically experience each day?
5. How much idle time do your workers/machines experience each day?
6. How is each workplace laid out (proximity of tools to worker)?
7. What are your actual setup versus standard setup times?
8. What are the machines' configurations?
9. What is the degree of knowledge of the machine by its operator?
10. How much scrap/rework do you experience each day?
11. What are your major bottleneck operations?

IV. Production Policies   *Response*

1. How flexible are your workers?
2. Do you cross-train?
3. Are your workers unionized?
4. Who sets the production standards?
5. How are workers evaluated?
6. How do you initiate production on new product introductions?
7. What is your relationship with suppliers?

V. Quality Control   *Response*

1. How do you measure material quality?
2. How do you measure end-item quality?
3. Do you have inspection points?
4. How do you account for human error/ variability?
5. Who sets the tolerances (specifications)?

General Comments:

Engineering Manager

I. Engineering Overview   *Response*

1. How complex are your products?
2. Do you have standard product designs?
3. To what degree are your parts standardized?
4. How well do you integrate activities in the engineering cycle?
5. What is your make-versus-buy policy?
6. Who sets this policy?
7. To what degree do you design for manufacturability?
8. How many engineering changes do you experience in a typical period?

General Comments:

Cost Accounting Manager

I. Cost Center Definition   *Response*

1. What are the major cost centers for production?
2. What are the major cost centers for each manufacturing process?
3. What are the major cost centers for product engineering?
4. What are the major cost centers for process engineering?
5. What are the major cost centers for quality control?
6. What depreciation methods do you use?
7. How do you determine what long-term projects/strategic investments to undertake?
8. What are your material storage and handling costs?

II. Cost Allocations–Production/   *Allocation*
   Operations   *Method*

1. Labor
   a. Touch
      1. Setup
      2. Load/unload
      3. Run
      4. Idle
      5. Scrap/rework
      6. Administrative and personal allowance
      7. Fringe
   b. Business activity
      1. Supervision
      2. Process labor
      3. Administrative and personal allowance
      4. Fringe

2. Material
  a. Product related
    1. Raw material or components
    2. Scrap/rework/yield
    3. Operating supplies (expendables)
    4. Cut off
    5. PPV
  b. Process related
    1. Operating supplies

3. Technology
  a. Machines and equipment (hardware and software)
  b. Finance (cost of capital)
  c. Hardware/software installation and test
  d. Technology acquisition and cost
  e. Unplanned capacity

III. Cost Allocation–Material Movement

Allocation Method

1. Product
2. Process
  a. WIP stores (EOQ)
  b. Material handling (people and technology)
  c. WIP cost of capital

IV. Cost Allocation–Incoming Material Control

Allocation Method

1. Procurement
  a. Nonproduction related
    1. Buyers
    2. Technology
    3. Management/administrative support
    4. Subcontractor mgmt.
    5. PPV
    6. Receiving

   b. Production related
    1. Buyers
    2. Technology
    3. Management/administrative support
    4. Subcontractor mgmt.
    5. PPV
    6. Receiving

2. Incoming Inspection
3. Vendor Qualifications
4. Vendor Performance
5. Raw Material Storage
   a. Cost of capital
   b. Technology
   c. People
6. Transport Material to First Operation

V. Cost Allocation–Outgoing Material Control

Allocation Method

1. Transport from Last Operation to Storage
2. Finished Goods Stores
   a. Cost of capital
   b. Technology
   c. People
3. Packaging/Handling/Shipping
4. Distribution/Transportation

VI. Cost Allocation–Facilities

Allocation Method

1. Maintenance
   a. Preventive
   b. Repair
2. Security and Emergency
3. Energy Management
4. Plant Rearrangement/Construction

VII. Cost Allocation–Quality Control

Allocation Method

1. Q/C Planning
2. Q/C Activities

VIII. Cost Allocation–Production Management

Allocation Method

1. Planning and Control Systems
2. Cell Planners and Dispatchers
3. Production Monitoring
4. Time and Attendance
5. Financial Control

IX. Cost Allocation–Process Dev./ Maintenance

Allocation Method

1. New Product or Major Changes
2. Maintenance
   a. Tooling and production programming
   b. Marketing
   c. Product service
   d. Human factors
   e. Information systems

General Comments:

Materials Manager

I. Management Overview

Response

1. What is your make-versus-buy policy?
2 What are your vendor capabilities?
3. How do you specify quality?
4. What are your performance measures?

II. Factors of Production   *Response*

    1. What are your resource limitations?
       a. Capital
       b. Labor pool
       c. Material supply
    2. How variable are your products?
    3. How accurate is the forecast?
    4. How accurate is the documentation?
       a. Routings
       b. Bills of materials
       c. Standards
    5. What is your average inventory on hand?

III. Supplier Management   *Response*

    1. How many vendors do you deal with?
    2. Where are they located?
    3. What methods of transportation do you use?
    4. What is your raw material lead time?
    5. What are your vendors' capabilities?
    6. How would you characterize communications with your vendors?
    7. How do you schedule deliveries?

IV. Inventory   *Response*

    1. What is your stocking policy?
    2. What are your average inventory levels?
       a. Finished goods
       b. RIP or WIP
       c. Component raw material

General Comments:

BIBLIOGRAPHY

Andrew, Charles G. "Peopleware: Driving the Productivity Machine." *P&IM Review and APICS News*, February 1985, pp. 36, 40, 42, 85.

Anthony, Robert. *Planning and Control Systems: A Framework for Analysis*. Boston, MA: Division of Research, Graduate School of Business Administration, Harvard University, 1965.

Anthony, Robert, and Regina Herzlinger. *Management Control in Nonprofit Organizations*. Homewood, IL: Richard D. Irwin, Inc., 1975.

Argyris, Chris. *Inner Contradictions of Rigorous Research*. New York: Academic Press, 1980.

Bennett, Robert E., and James A. Hendricks. "Justifying the Acquisition of Automated Equipment." *Management Accounting*, July 1987, pp. 39–46.

Berry, William D., and Michael S. Lewis-Beck, editors. *New Tools for Social Scientists: Advances and Applications in Research Methods*. Beverly Hills, CA: Sage Publications, 1986.

Berry, William D.; Michael S. Lewis-Beck; David E. Keys; and Edward J. Rudnicki. *Cost Accounting for Factory Automation*. Montvale, NJ: The National Association of Accountants, 1987.

Brandt, Richard, and Otis Port. "How Automation Could Save the Day." *Business Week*, March 3, 1986, pp. 72–74.

Brimson, James A. "How Advanced Manufacturing Technologies are Reshaping Cost Management." *Management Accounting*, March 1986, pp. 25–29.

————. "Computer-Integrated Manufacturing: Vision or Illusion?" *Business Software Review*, April 1987, pp. 42–47.

Church, A. Hamilton. *Manufacturing Costs and Accounts*. New York: McGraw-Hill Book Company, Inc., 1929.

Cooper, Robin. "Cost Management Concepts and Principles." *Journal of Cost Management*, Summer 1987, pp. 43–51.

————. "Cost Management Concepts and Principles." *Journal of Cost Management*, Spring 1987, pp. 45–49.

Cooper, Robin, and Robert S. Kaplan. "How Cost Accounting Systematically Distorts Product Costs." Working Paper—Harvard University Graduate School of Business, 1986.

Covaleski, M. A., and M. W. Dirsmith. "Budgeting as a Means for Control and Loose Coupling." *Accounting, Organizations and Society,* 1983, pp. 323–340.

Cushing, Barry. "On the Feasibility and the Consequences of a Database Approach to Corporate Financial Reporting." Working Paper, The Pennsylvania State University, April 1987.

Deal, Terrance E., and Allen A. Kennedy. *Corporate Cultures: The Rites and Rituals of Corporate Life.* Reading, MA: Addison-Wesley Publishing Corp., 1982.

Dilts, David M. and Grant W. Russell. "Accounting for the Factory of the Future." *Management Accounting,* April 1985, pp. 34–40.

Dray, Susan M. "Avoiding the Big, Bad Wolf: Proactively Managing the 'People Issues' in Technology Introduction." Presentation made at The Automation Forum, The Pennsylvania State University, February 1987.

Eccles, Robert G. "Analyzing Your Company's Transfer Pricing Practices." *Journal of Cost Management,* Summer, 1987, pp. 21–33.

Foster, George, and C. T. Horngren. "JIT: Cost Accounting and Cost Management Issues." *Management Accounting,* June 1987, pp. 19–25.

Fox, Robert E. "Main Bottleneck on the Factory Floor?" *Management Review,* November 1984, pp. 55–61.

_____. "Cost Accounting: Asset or Liability?" *Journal of Accounting and EDP,* Winter 1986, pp. 31–37.

_____. "Coping with Today's Technology: Is Cost Accounting Keeping Up?" In *Cost Accounting for the 90s: The Challenge of Technological Change, Proceedings.* Montvale, NJ: National Association of Accountants 1986, pp. 13–26.

Glaser, Barney G., and Anselm Strauss. *The Discovery of Grounded Theory: Strategies for Qualitative Research.* New York: Aldine Publishing Company, 1967.

Haas, Elizabeth A. "Breakthrough Manufacturing." *Harvard Business Review,* March–April 1987, pp. 75–81.

Haley, Roy W., and Bruce B. Piper. "New Inventory-Management Approach Can Substantially Cut Inventory Costs." *The Practical Accountant,* October 1986, pp. 60–68.

Hall, Robert W. *Attaining Manufacturing Excellence.* Homewood, IL: Dow Jones-Irwin, 1987.

Harper, Chris. "Zero-based Inventories Get JIT Working." *Paperboard Packaging,* May 1987, pp. 44, 46, 48.

Hayes, Robert, and W. J. Abernathy. "Managing Our Way to Economic Decline." *Harvard Business Review* 58: July–August 1980, pp. 67–77.

Hayes, Robert, and Steven C. Wheelwright. *Restoring Our Competitive Edge: Competing Through Manufacturing.* New York: John Wiley & Sons, 1984.

Holbrook, William. "Practical Accounting Advice for Just-in-Time Production." *Journal of Accounting and EDP,* Fall 1985, pp. 42–47.

Hopwood, A. G. "The Archeology of Accounting Systems." *Accounting, Organizations and Society* 12, no. 3, 1987, pp. 207–234.

―――. "On Trying to Study Accounting in the Contexts in Which It Operates." *Accounting, Organizations and Society,* 1983, pp. 287–305.

―――. *Accounting and Human Behavior.* Englewood Cliffs, NJ: Prentice-Hall, Inc., 1974.

Howell, Robert A. "Changing Measurements in the Factory of the Future." In *Cost Accounting for the 90s: The Challenge of Technological Change, Proceedings.* Montvale, NJ: National Association of Accountants, 1986, pp. 105–116.

Howell, Robert A.; James D. Brown; Stephen Soucy; and Allen Seed III. *Management Accounting in the New Manufacturing Environment: Current Cost Management Practice in Automated (Advanced Manufacturing Environments).* Montvale, NJ: The National Association of Accountants, 1987.

Howell, Robert A., and Stephen R. Soucy. "The New Manufacturing Environment: Major Trends for Management Accountants." *Management Accounting,* July 1987, pp. 21–27.

―――. "Cost Accounting in the New Manufacturing Environment." *Management Accounting,* August 1987, pp. 42–49.

Jayson, Susan. "Goldratt and Fox: Revolutionizing the Factory Floor." *Management Accounting,* May 1987, pp. 18–25.

Johansson, Hank. "The Revolution in Cost Accounting." *P&IM Review and APICS News,* January 1985, pp. 42–46.

Johansson, Hank; Thomas E. Vollman; and Vivian Wright. "The Effect of Zero Inventories on Cost (Just-in-Time)." In *Cost Accounting for the 90s: The Challenge of Technological Change, Proceedings.* Montvale, NJ: National Association of Accountants, 1986, pp. 141–164.

Johnson, H. Thomas. "The Decline of Cost Management: A Reinterpretation of 20th Century Cost Accounting." *Journal of Cost Management,* Spring 1987, pp. 5–12.

Johnson, H. Thomas, and Dennis A. Loewe. "How Weyerhaeuser Manages Corporate Overhead Costs." *Management Accounting,* August 1987, pp. 20–26.

Johnson, H. Thomas, and Robert S. Kaplan. *Relevance Lost: The Rise and Fall of Management Accounting.* Boston, MA: Harvard Business School Press, 1987.

Johnson, H. Thomas. "The Rise and Fall of Management Accounting." *Management Accounting,* January 1987, pp. 22–31.

Jordan, Henry H. "The Challenge of Implementing a Zero Inventories Program." *P&IM Review and APICS News,* June 1985, pp. 70, 85–86.

Kaplan, Robert S. "Strategic Cost Analysis." In *Cost Accounting for the 90s: The Challenge of Technological Change, Proceedings.* Montvale, NJ: National Association of Accountants, 1986, pp. 129–140.

_____. "Measuring Manufacturing Performance: A New Challenge for Managerial Accounting Research." *The Accounting Review* 58, no. 4, 1983, pp. 686–705.

Kuhn, Thomas. *The Structure of Scientific Revolutions.* Chicago: University of Chicago Press, 1957.

Keys, David E. "Six Problems in Accounting for N/C Machines." *Management Accounting,* November 1986, pp. 38–47.

Lammert, Thomas B., and Robert Ehrsam. "The Human Element: The Real Challenge in Modernizing Cost Systems." *Management Accounting,* July 1987, pp. 32–38.

Lincoln, Yvonna S., and Egon G. Guba. *Naturalistic Inquiry.* Beverly Hills, CA: Sage Publications, 1985.

Littrell, Earl K. "The High Tech Challenge to Management Accounting." *Management Accounting,* October 1984, pp. 33–36.

Mackey, Jim. "Key Issues in Manufacturing Accounting." *Management Accounting,* January 1987, pp. 32–38.

Major, Michael J. "The Bottom Line in Manufacturing." *Business Software Review,* August 1986, pp. 51–53, 57.

McGinty, Patrick. "Cost Accounting Revisited: A Manufacturing Survey." *Journal of Cost Management,* Spring 1987, pp. 35–44.

McIlhatten, Robert D. "The Path to Total Cost Management." *Journal of Cost Management,* Summer 1987, pp. 5–10.

_____. "How Cost Management Systems Can Support the JIT Philosophy." *Management Accounting,* September 1987, pp. 20–27.

McKinnon, William P., and Ernest A. Kallman. "Mapping Chargeback Systems to Organizational Environments." *MIS Quarterly,* March 1987, pp. 5–19.

McClintock, Charles C.; Diane Brannon; and Steven Maynard-Moody. "Applying the Logic of Sample Surveys to Qualitative Case Studies: The Case Cluster Method." *Administrative Science Quarterly,* December 1979, pp. 612–629.

McNair, C. J., and William Mosconi. "Measuring Performance in an Advanced Manufacturing Environment." *Management Accounting,* July 1987, pp. 28–31.

Merchant, Kenneth A. *Control in Business Organizations.* Boston: Pitman Publishing Company, 1985.

Mitroff, Ian I., and Susan A. Mohrman. "The Slack Is Gone: How the United States Lost Its Competitive Edge in the World Economy." *Academy of Management Executive,* February 1987, pp. 65–70.

Morse, Wayne J.; Harold P. Roth; and Kay M. Poston. *Measuring, Planning and Controlling Quality Costs.* Montvale, NJ: The National Association of Accountants, 1987.

Neumann, Bruce R., and Pauline R. Jaouen. "Kanban, ZIPS, and Cost Accounting: A Case Study." *Journal of Accountancy,* August 1986, pp. 132–142.

Nolan, Richard. "Corporate Database: The Time is Now." *Harvard Business Review,* 1973.

———. "Controlling the Costs of Data Services." *Harvard Business Review,* July–August 1977.

Peters, Thomas J., and Robert H. Waterman, Jr. *In Search of Excellence: Lessons from America's Best-Run Companies.* New York: Harper & Row, Publishers, 1982.

Primrose, P. L., and R. Leonard. "Performing Investment Appraisals for Advanced Manufacturing Technology." *Journal of Cost Management,* Summer 1987, pp. 34–42.

Rae, Sharon Gamble. "How CIM Will Retool Manufacturing—Someday." *ICP Manufacturing Software,* Autumn 1985, pp. 11–16.

Richardson, Peter R. "Managing Costs Strategically." *Journal of Cost Management,* Summer 1987, pp. 11–20.

Rohan, Thomas M. "Justifying Your CIM Investment." *Industry Week,* March 9, 1987, pp. 33–35.

Russell, Grant W., and David M. Dilts. "Are Accountants Delaying the Automation of America?" In *Cost Accounting for the 90s: The Challenge of Technological Change, Proceedings.* Montvale, NJ: National Association of Accountants, 1986, pp. 27–44.

Schonberger, R. *World Class Manufacturing: The Lessons of Simplicity Applied.* New York: The Free Press, 1986.

Seed, Allen. "But How Do You Cost-Justify It?" *Computerworld,* 1986, pp. 28–30.

———. "Cost Accounting in the Age of Robotics." *Management Accounting,* October 1984, pp. 39–43.

Sepehri, Mehran. *Just-in-Time, Not Just in Japan.* Falls Church, VA: American Production and Inventory Control Society, Inc., 1986.

Shillinglaw, Gordon. *Managerial Cost Accounting.* Fifth Edition. Homewood, IL: Richard D. Irwin, Inc., 1982.

Simpson, James B., and David L. Muthler. "Quality Costs: Facilitating the Quality Initiative." *Journal of Cost Management,* Spring 1987, pp. 25–34.

Sorter, George. "An 'Events' Approach to Basic Accounting Theory." *Accounting Review* 44, no. 1, 1969, pp. 12–20.

Susman, Gerald I., and James W. Dean, Jr. "Strategic Use of Advanced Manufacturing Technology in the Emerging Competitive Environment." Working Paper—Center for the Management of Technological and Organizational Change." The Pennsylvania State University, January 1987.

Tatikonda, Lakshmi. "Production Managers Need a Course in Cost Accounting." *Management Accounting,* June 1987, pp. 26–29.

Turney, Peter B. B., and Bruce Anderson. "Cost Accounting Joins the Manufacturing Revival." Working Paper, April 1987.

Vangermeersch, Richard. "Renewing Our Heritage." *Management Accounting,* July 1987, pp. 47–49.

————. "Milestones in the History of Management Accounting." In *Cost Accounting for the 90s: The Challenge of Technological Change* Proceedings. Montvale, NJ: National Association of Accountants, 1986, pp. 75–84.

Van Maanen, John; James M. Dabbs, Jr.; Robert R. Faulkner. *Varieties of Qualitative Research.* Beverly Hills, CA: Sage Publications, 1987.

Vollum, Robert B. "Cost Accounting: The Key to Capturing Cost Information on the Factory Floor." *Journal of Accounting and EDP,* Summer 1985, pp. 44–51.

Walton, Richard E., and Gerald I. Susman. "People Policies for the New Machines." *Harvard Business Review,* March–April 1987, pp. 98–106.

Worthy, Ford S. "Accounting Bores You? Wake Up." *Fortune,* October 12, 1987, pp. 43–44, 48–50.

INDEX

A

Abernathy, W. J., 196
Academia, shortcomings of, and
  cost accounting, 50–51
Accountability, individual, 65–66
Accountants
  and basics of accounting, 66–67
  common view of, 64–65
  at CompSci Industries, 73–77
  concept of as rigid, 65
  at Diesel Systems, Inc., 72–73
  failure to change of, 67–68
  at General Business, Inc., 77–79
  at MicroChip, Inc., 69–72
  proactive; *see* Proactive
    accountants
  reactive, 67
  at TeleComm Corporation, 67–68
Accounting; *see also* Management
  accounting system (MAS)
  centralization of, 121–23
  cost, 181–83
  JIT, 143–62
  matched to cost flow, 74–75
  responsibility, 65–66, 113, 114–
    16, 126
  simplified, 129–35, 140–41, 146–
    47, 148
  strategy reflection by, 147
Accounting numbers, impact on behavior of, 176–77
Activity costing, 111

Actual costs
  continuous improvement sup-
    ported by, 131–32
  evaluation system and, 132
  replacement of standards with,
    82–85, 131–32
  as starting point, 132–33
  universal adoption of, 85–86
Advanced manufacturing technolo-
    gies (AMT), 2–3; *see also* Tech-
    nology *and* Technology
    adoption
  accounting change forced by,
    127–29
  capital budgeting and, 165–70
  cost pattern changes associated
    with, 51–52
  financial controls in, 14
  and flexibility, 13
  key trends resulting from 14–15
  migration path of, 44–45
  problem amplification by, 4–5
  standardization requirement of,
    93
  strategic implementation of, 45–
    46
  various types of, 15–17; *see also*
    specific type
Allocation, 98–99
  apportionment versus, 97, 110
  material value and, 105
Allocation techniques, 183